SpringerBriefs in Optimization

Series Editors

Panos M. Pardalos
János D. Pintér
Stephen M. Robinson
Tamás Terlaky
My T. Thai

SpringerBriefs in Optimization showcases algorithmic and theoretical techniques, case studies, and applications within the broad-based field of optimization. Manuscripts related to the ever-growing applications of optimization in applied mathematics, engineering, medicine, economics, and other applied sciences are encouraged.

For further volumes:
http://www.springer.com/series/8918

Slawomir Koziel • Stanislav Ogurtsov

Antenna Design by Simulation-Driven Optimization

 Springer

Slawomir Koziel
School of Science and Engineering
Reykjavik University Engineering
    Optimization & Modeling Center
Reykjavik, Iceland

Stanislav Ogurtsov
School of Science and Engineering
Reykjavik University Engineering
    Optimization & Modeling Center
Reykjavik, Iceland

ISSN 2190-8354   ISSN 2191-575X (electronic)
ISBN 978-3-319-04366-1   ISBN 978-3-319-04367-8 (eBook)
DOI 10.1007/978-3-319-04367-8
Springer Cham Heidelberg New York Dordrecht London

Library of Congress Control Number: 2013958383

Printed on acid-free paper

Springer is part of Springer Science+Business Media (www.springer.com)

Preface

Design of contemporary antenna structures heavily relies on electromagnetic (EM) simulations. Accurate reflection and radiation responses of many antenna geometries can be obtained only with discrete full-wave EM simulation. On the other hand, the direct use of high-fidelity EM simulation in the design process, particularly for automated parameter optimization, results in high computational costs, often prohibitive. Other issues, such as the presence of numerical noise, may result in a failure of optimization using conventional (e.g., gradient-based) methods. In this book, we demonstrate that numerically efficient design of antennas can be realized using surrogate-based optimization (SBO) methodologies. The essence of SBO techniques resides in shifting the optimization burden to a fast surrogate of the antenna structure and the use of coarse-discretization EM models to configure the surrogate. A properly created and handled surrogate serves as a reliable prediction tool so that satisfactory designs can be found at the costs of a limited number of simulations of the high-fidelity EM antenna model. The specific SBO techniques covered here include space mapping combined with response surface approximation, shape-preserving response prediction (SPRP), adaptive response correction (ARC), adaptively adjusted design specification (AADS), variable-fidelity simulation-driven optimization (VFSDO), and surrogate-based optimization enhanced by the use of adjoint sensitivities. Multi-objective design of antennas is also covered to some extent. Moreover, we discuss practical issues such as the effect of the coarse-discretization model fidelity on the final design quality and the computational cost of the optimization process. Our considerations are illustrated using numerous application examples. Recommendations concerning application of specific SBO techniques to antenna design are also presented.

Contents

Chapter 1
Introduction

Design of modern antennas is undoubtedly a challenging task. An important part of the design process is the adjustment of geometry and material parameters to ensure that the antenna response satisfies prescribed performance specifications with respect to certain characteristics such as input impedance, radiation pattern, antenna efficiency, etc. (Volakis 2007; Schantz 2005; Petosa 2007; Balanis 2005). In this context, computationally inexpensive analytical models can only be used—in most cases—to obtain an initial estimate of the optimum design. This is particularly the case when certain interactions within the antenna itself and with the antenna environment (e.g., housing, installation fixture, feeding circuit, connectors) have to be taken into account. For these reasons, full-wave electromagnetic (EM) simulation plays an essential role in the design closure. Contemporary computational techniques—implemented in commercial simulation packages—are capable to adequately evaluate antenna reflection and radiation responses. On the other hand, full-wave simulations of realistic and finely discretized antenna models are computationally expensive: evaluation for a single combination of design parameters may take up to several hours. While this cost is acceptable from the design validation standpoint, it is usually prohibitive for design optimization that normally requires a large number of EM simulations of the antenna structure of interest.

Automation of the antenna design process can be realized by formulating the antenna parameter adjustment task as an optimization problem with the objective function supplied by an EM solver (Special issue, IEEE APS 2007). Unfortunately, most of the conventional optimization techniques, including gradient-based (Nocedal and Wright 2000), e.g., conjugate-gradient, quasi-Newton, sequential quadratic programming, etc., and derivative-free (Kolda et al. 2003) methods, e.g., Nelder-Mead and pattern search techniques, require a large number of objective function evaluations to converge to a satisfactory design. For many realistic EM antenna models, where evaluation time per design reaches a few hours with contemporary computing facilities, the cost of such an optimization process may be unacceptably high. Another practical problem of conventional optimization techniques is numerical noise, which is partially a result of adaptive meshing techniques used

S. Koziel and S. Ogurtsov, *Antenna Design by Simulation-Driven Optimization*,
SpringerBriefs in Optimization, DOI 10.1007/978-3-319-04367-8_1,
© Slawomir Koziel and Stanislav Ogurtsov 2014

by most contemporary EM solvers: even a small change of design variables may result in a change of the mesh topology and, consequently, discontinuity of the EM-simulated antenna responses as a function of designable parameters. The noise is particularly an issue for gradient-based methods that normally require smoothness of the objective function.

The aforementioned challenges result in a situation where the most common approach to simulation-driven antenna design is based on repetitive parameter sweeps (usually, one parameter at a time). This approach is usually more reliable than a brute-force optimization using built-in optimization capabilities of commercial simulation tools; however, it is also very laborious, time-consuming, and demanding significant designer supervision. Moreover, such a parameter-sweep-based optimization process does not guarantee optimum results because only a limited number of parameters can be handled that way. It is also difficult to utilize correlations between the parameters properly. Finally, optimal values of the designable variables can be quite counterintuitive.

In recent years, population-based search methods (also referred to as metaheuristics) (Yang 2010) have gained considerable popularity. This group of methods includes, among others, genetic algorithms (GA) (Back et al. 2000), particle swarm optimizers (PSO) (Kennedy 1997), differential evolution (DE) (Storn and Price 1997), and ant colony optimization (Dorigo and Gambardella 1997). Most of metaheuristics are biologically inspired systems designed to alleviate certain difficulties of the conventional optimization methods, in particular, handling problems with multiple local optima (Yang 2010). Probably the most successful application of the metaheuristic algorithms in antenna design resided so far in array optimization problems (e.g., Ares-Pena et al. 1999; Haupt 2007; Jin and Rahmat-Samii 2007, 2008; Petko and Werner 2007; Bevelacqua and Balanis 2007; Grimaccia et al. 2007; Pantoja et al. 2007; Selleri et al. 2008; Li et al. 2008; Rajo-Iglesias and Quevedo-Teruel 2007; Roy et al. 2011). In these problems, the cost of evaluating the single element response is not of the primary concern or the response of a single element is already available, e.g., with a preassigned array element. However, application of metaheuristics to EM-simulation-driven antenna design is not practical because corresponding computational costs would be tremendous: typical GA, PSO, or DE algorithm requires hundreds, thousands, or even tens of thousands of objective function evaluations to yield a solution (Ares-Pena et al. 1999; Haupt 2007; Jin and Rahmat-Samii 2007, 2008; Petko and Werner 2007; Bevelacqua and Balanis 2007; Grimaccia et al. 2007; Pantoja et al. 2007; Selleri et al. 2008; Li et al. 2008; Rajo-Iglesias and Quevedo-Teruel 2007; Roy et al. 2011).

The problem of high computational cost of conventional EM-based antenna optimization can be alleviated to some extent by the use of adjoint sensitivity (Director and Rohrer 1969), which is a computationally cheap way to obtain derivatives of the system response with respect to its design parameters. Adjoint sensitivities can substantially speed up microwave design optimization while using gradient-based algorithms (Bandler and Seviora 1972; Chung et al. 2001). This technology was also demonstrated for antenna optimization (Jacobson and Rylander 2010; Toivanen et al. 2009; Zhang et al. 2012). It should be mentioned, however, that adjoint sensitivities are not yet widespread in commercial EM solvers. Only CST (CST Microwave Studio 2011) and HFSS (HFSS 2010) have recently implemented this feature. Also, the use of adjoint sensitivities is limited by numerical noise of the EM-simulated response (Koziel et al. 2012c).

One of the most recent and yet most promising ways to realize computationally efficient simulation-driven antenna design is surrogate-based optimization (SBO) (Koziel and Ogurtsov 2011a, b; Forrester and Keane 2009; Queipo et al. 2005). The SBO main idea is to shift the computational burden of the optimization process to a surrogate model which is a cheap representation of the optimized antenna (Bandler et al. 2004a, b; Queipo et al. 2005; Koziel et al. 2006; Koziel and Ogurtsov 2011a). In a typical setting, the surrogate model is used as a prediction tool to find approximate location of the original (high-fidelity or fine) antenna model. After evaluating the high-fidelity model at this predicted optimum, the surrogate is updated in order to improve its local accuracy (Koziel et al. 2011c). The key prerequisite of the SBO paradigm is that the surrogate is much faster than the high-fidelity model. Also, in many SBO algorithms, the high-fidelity model is only evaluated once per iteration. Therefore, the computational cost of the design process with a well working SBO algorithm may be significantly lower than those with most of conventional optimization methods.

There are two major types of surrogate models. The first one comprises function-approximation models constructed from sampled high-fidelity simulation data (Simpson et al. 2001). A number of approximation (and interpolation) techniques are available, including artificial neural networks (Haykin 1998), radial basis functions (Gutmann 2001; Wild et al. 2008), kriging (Forrester and Keane 2009), support vector machines (Smola and Schölkopf 2004), Gaussian process regression (Angiulli et al. 2007; Jacobs 2012), or multidimensional rational approximation (Shaker et al. 2009). If the design space is sampled with sufficient density, the resulting model becomes reliable so that the optimal antenna design can be found just by optimizing the surrogate. In fact, approximation methods are usually used to create multiple-use library models of specific components. The computational overhead related to such models may be very high. Depending on the number of designable parameters, the number of training samples necessary to ensure decent accuracy might be hundreds, thousands, or even tens of thousands. Moreover, the number of samples quickly grows with the dimensionality of the problem (so-called curse of dimensionality). As a consequence, globally accurate approximation modeling is not suitable for ad hoc (one-time) antenna optimization. Iteratively improved approximation surrogates are becoming popular for global optimization (Couckuyt 2013). Various ways of incorporating new training points into the model (so-called infill criteria) exist, including exploitative models (i.e., models oriented toward improving the design in the vicinity of the current one), explorative models (i.e., models aiming at improving global accuracy), as well as model with balanced exploration and exploitation (Jones et al. 1998; Forrester and Keane 2009).

Another type of surrogates, so-called physics-based surrogates, is constructed from underlying low-fidelity (or coarse) models or the respective structures. Because the low-fidelity models inherit some knowledge of the system under consideration, usually a small number of high-fidelity simulations are sufficient to configure a reliable surrogate. The most popular SBO approaches using physics-based surrogates that proved to be successful in microwave engineering are space mapping (SM) (Bandler et al. 2004a, b), tuning, tuning SM (Koziel et al. 2009a; Cheng et al. 2010), as well as various response correction methods (Echeverria and Hemker 2005; Koziel et al. 2009b; Koziel 2010a). To ensure computational efficiency, it is important that the low-fidelity model is considerably faster than the high-fidelity model. For that reason, circuit equivalents or models based on analytical formulas are preferred (Bandler et al. 2004a, b). The aforementioned methods (particularly space mapping) were mostly used to design filters or transmission-line-based components such as impedance transformers (Amari et al. 2006; Wu et al. 2004; Bandler et al. 2004a, b). Unfortunately, in case of antennas, reliable circuit equivalents are rarely available. For antennas, a universal way of obtaining low-fidelity models is through coarse-discretization EM simulations. Such models are relatively expensive, which poses additional challenges in terms of optimization.

The topic of this book is surrogate-based optimization methods for simulation-driven antenna design with the focus on surrogate-based techniques exploiting variable-fidelity EM simulations and physics-based surrogates. We begin, in Chap. 2, by formulating the antenna design task as an optimization problem. We also briefly discuss conventional numerical optimization techniques, including both gradient-based and derivative-free methods but also metaheuristics. In Chap. 3, surrogate-based optimization is introduced. In the same chapter, the SBO design workflow as well as various aspects of surrogate-based optimization is presented on a generic level. Chapter 4 is an exposition of the specific state-of-the-art physics-based SBO techniques that are suitable for antenna design optimization. The emphasis is put on methods that aim at minimizing the number of both high- and low-fidelity EM simulations of the antenna under design and thus reducing the overall design cost. Chapters 6–9 present applications of the methods discussed in Chap. 4 for the design of specific antenna structures. Variable-fidelity design exploiting adjoint sensitivity is presented in Chap. 10. Chapter 11 discusses multi-objective antenna design using surrogate models. Chapter 12 provides a discussion of open issues related to SBO antenna design with special focus on selecting simulation model fidelity and its impact on the performance and computational efficiency of the optimization process. The book is concluded in Chap. 13. Here, we formulate recommendations for the readers interested in applying presented algorithm and techniques in their antenna design and discuss possible future developments concerning mostly automation of simulation-driven antenna design.

Chapter 2
Antenna Design Using Electromagnetic Simulations

In this chapter, we formulate the antenna design task as a nonlinear minimization problem. We introduce necessary notation, discuss typical objectives and constraints, and give a brief overview of conventional optimization techniques, including gradient-based and derivative-free methods, as well as metaheuristics. We also introduce the concept of the surrogate-based optimization (SBO) and discuss it on a generic level. More detailed exposition of SBO and SBO-related design techniques will be given in Chaps. 3 and 4.

2.1 Antenna Design Task as an Optimization Problem

Let $R_f(x)$ denote a response of a high-fidelity (or fine) model of the antenna under design. For the rest of this book, we will assume that R_f is obtained using accurate full-wave electromagnetic (EM) simulation. Typically, R_f will represent evaluation of performance characteristics, e.g., reflection $|S_{11}|$ or gain over certain frequency band of interest. Vector $x = [x_1\ x_2...x_n]^T$ represents designable parameters to be adjusted (e.g., geometry and/or material ones).

In some situations, individual components of the vector $R_f(x)$ will be considered, and we will use the notation $R_f(x) = [R_f(x, f_1)\ R_f(x, f_2)\ ...\ R_f(x, f_m)]^T$, where $R_f(x, f_k)$ is the evaluation of the high-fidelity model at a frequency f_k, whereas f_1 through f_m represent the entire discrete set of frequencies at which the model is evaluated.

The antenna design task can be formulated as the following nonlinear minimization problem (Koziel and Ogurtsov 2011a):

$$x^* = \arg\min_{x} U\left(R_f(x)\right) \tag{2.1}$$

where U is the scalar merit function encoding the design specifications, whereas x^* is the optimum design to be found. The composition $U(R_f(x))$ will be referred to as the objective function. The function U is implemented so that a better design x corresponds to a smaller value of $U(R_f(x))$. In antenna design, U is most often

S. Koziel and S. Ogurtsov, *Antenna Design by Simulation-Driven Optimization*, SpringerBriefs in Optimization, DOI 10.1007/978-3-319-04367-8_2, © Slawomir Koziel and Stanislav Ogurtsov 2014

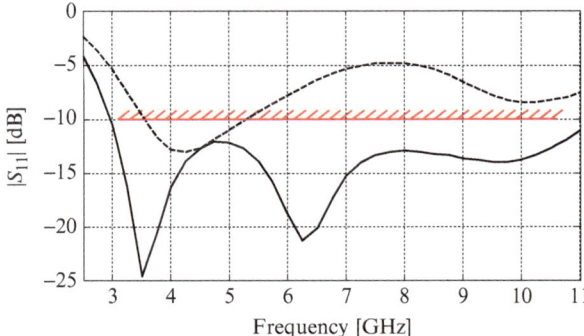

Fig. 2.1 Illustration of minimax design specifications, here, $|S_{11}| \leq -10$ dB for 3.1–10.6 GHz, marked with *thick horizontal line*. An example UWB antenna reflection response that does not satisfy our specifications (*dashed line*) (specification error, i.e., maximum violation of the requirements is about +5 dB) and another response that does satisfy the specifications (*solid line*)

implemented as a minimax function with upper (and/or lower) specifications. Figure 2.1 shows the example of minimax specifications corresponding to typical UWB requirements for the reflection response, i.e., $|S_{11}| \leq -10$ dB for 3.1–10.6 GHz. The value of $U(\mathbf{R}_f(\mathbf{x}))$ (also referred to as minimax specification error) corresponds to the maximum violation of the design specifications within the frequency band of interest.

To simplify notation, we will occasionally use the symbol $f(\mathbf{x})$ as an abbreviation for $U(\mathbf{R}_f(\mathbf{x}))$.

In reality, the problem (2.1) is always constrained. The following types of constraints can be considered:

- Lower and upper bounds for design variables, i.e., $l_k \leq x_k \leq u_k$, $k = 1, \ldots, n$
- Inequality constraints, i.e., $c_{\text{ineq.}l}(\mathbf{x}) \leq 0$, $l = 1, \ldots, N_{\text{ineq}}$, where N_{ineq} is the number of constraints
- Equality constraints, i.e., $c_{\text{eq.}l}(\mathbf{x}) = 0$, $l = 1, \ldots, N_{\text{eq}}$, where N_{eq} is the number of constraints

Design constraints are usually introduced to make sure that the antenna structure that is to be evaluated by the EM solver is physically valid (e.g., certain parts of the structure do not overlap). Also, constraints can be introduced in order to ensure that the physical dimensions (length, width, area) of the antenna do not exceed certain assumed values.

In this book, geometry constraints such as those described above are handled explicitly. Other types of constraints, particularly those that emerge due to converting initially multi-objective design problem into single-objective one, are handled through penalty functions. It should be mentioned though that the literature offers efficient ways of explicit handling expensive constraints; see, e.g., Kazemi et al. (2011), Basudhar et al. (2012).

Figure 2.2 shows the simulation-driven design optimization flowchart. Typically, it is an iterative process where the designs found by the optimizer are verified by

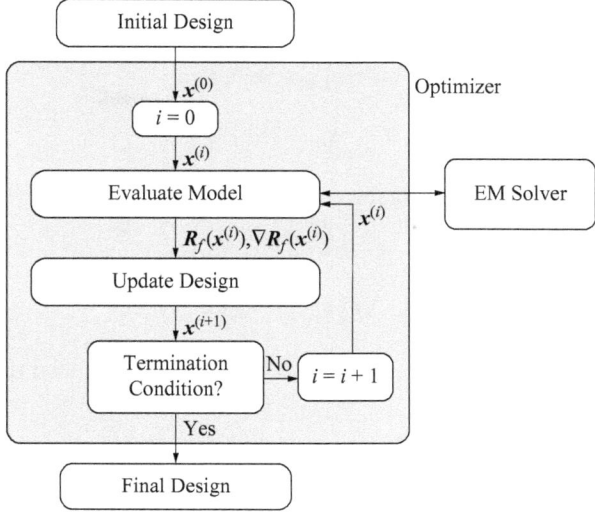

Fig. 2.2 Simulation-driven design through optimization. Generic optimization scheme is an iterative process where the new candidate designs are generated by the optimization algorithm and the high-fidelity model is evaluated through EM simulation for verification purposes and to provide the optimizer with information that allows searching for possible better designs. Depending on the type of the algorithm, the search process may be guided by the model response or (if available) by the response and its derivatives (gradient)

evaluating the high-fidelity model in the EM solver and—depending on a particular algorithm—the search process is guided either by the model response itself or the response of its gradients (if available). In Sects. 2.2–2.5, we briefly discuss conventional optimization approaches. In Chaps. 3 and 4, we discuss surrogate-based optimization methods, which are the main topic of this book.

2.2   Gradient-Based Optimization Methods

Gradient-based optimization techniques are the oldest and the most popular optimization methods (Nocedal and Wright 2000). In order to proceed toward the optimum design, they utilize derivative information of the objective function. Assuming that the objective $f(x)$ is sufficiently smooth (i.e., at least continuously differentiable), the gradient $\nabla f = [\partial f/\partial x_1 \; \partial f/\partial x_2 \ldots \partial f/\partial x_n]^\mathrm{T}$ gives the information about descent of f in the vicinity of the design at which the gradient is calculated. More specifically,

$$f(x+h) \cong f(x) + \nabla f(x) \cdot h < f(x) \tag{2.2}$$

for sufficiently small h as long as $\nabla f(x) \cdot h < 0$. In particular $h = -\nabla f(x)$ determines the direction of the steepest descent. In practice, using steepest descent as a search direction results in a poor performance of the optimization algorithm (Nocedal and

Wright 2000; Yang 2010). Better results are obtained using so-called conjugate-gradient method where the search direction is determined as a combination of the previous direction h_{prev} and the current gradient, i.e.,

$$h = -\nabla f\left(x^i\right) + \gamma h_{prev} \tag{2.3}$$

An example way of selecting the coefficient γ is a Fletcher-Reeves method with

$$\gamma = \frac{\nabla f\left(x\right)^{\mathrm{T}} \nabla f\left(x\right)}{\nabla f\left(x_{prev}\right)^{\mathrm{T}} \nabla f\left(x_{prev}\right)} \tag{2.4}$$

Having the search direction, the next design x^{i+1} is determined from the current one x^i as

$$x^{i+1} = x^i + \alpha \cdot h \tag{2.5}$$

Here, the choice of the step size $\alpha > 0$ is of great importance (Nocedal and Wright 2000), and finding it is referred to as a line search.

It is also possible to utilize second-order derivative information, which is characteristic to so-called Newton methods. Assuming f is at least twice continuously differentiable, one can consider a second-order Taylor expansion of f:

$$f\left(x+h\right) \cong f\left(x\right) + \nabla f\left(x\right) \cdot h + \frac{1}{2} h \cdot H\left(x\right) \cdot h \tag{2.6}$$

where $H(x)$ is the Hessian of f at x, i.e., $H(x) = [\partial^2 f / \partial x_j \partial x_k]_{j,k=1,\ldots,n}$. This means, given the current design x^i being sufficiently close to the minimum of f, that the next approximation of the optimum can be determined as

$$x^{i+1} = x^i - \left[H\left(x\right)\right]^{-1} \nabla f\left(x\right) \tag{2.7}$$

If the starting point is sufficiently close to the optimum and the Hessian is positive definite (Yang 2010), the algorithm (2.7) converges very quickly to the (locally) optimal design. In practice, neither of these conditions is usually satisfied, so various types of damped Newton techniques are used, e.g., Levenberg-Marquardt method (Nocedal and Wright 2000). On the other hand, the Hessian of the objective function f is normally not available so quasi-Newton methods are used instead where the Hessian is approximated using various updating formulas (Nocedal and Wright 2000).

From the point of view of simulation-driven antenna design, the use of gradient-based methods is problematic mostly because of the high computational cost of accurate simulation and the fact that gradient-based methods normally require large number of objective function evaluations to converge, unless cheap way of obtaining sensitivity is utilized (e.g., through adjoints or automatic differentiation). Another problem is numerical noise that is always present in EM-based objective functions. Recently, adjoint sensitivity techniques have become available in some commercial EM solvers (CST 2013; HFSS 2010), which may revive the interest in this type of methods for antenna design because they allow calculation of sensitivity at little or no extra cost compared to a regular EM simulation of the antenna structure. On the other hand, automatic differentiation is usually not an option because source codes are not accessible whenever commercial solvers are utilized. It should also be mentioned that gradient-based methods exploiting a trust-region framework are usually more efficient than those based, e.g., on line search so that using trust region (Conn et al. 2000) is recommended whenever possible.

2.3   Derivative-Free Optimization Methods

In many situations, gradient-based search may not be a good option. This is particularly the case when derivative information is not available or expensive to compute (e.g., through finite differentiation of an expensive objective function). Also, if the objective function is noisy (which is typical for responses obtained from EM simulation) then the gradient-based search does not perform well.

Optimization techniques that do not use derivative data in the search process are referred to as derivative-free methods. Formally speaking metaheuristics (Sect. 2.4) as well as many surrogate-based approaches (Chaps. 3 and 4) also fall into this category. In this section, however, we only mention the basic idea of the local search methods. Figure 2.3 illustrates the concept of the pattern search (Kolda et al. 2003), where the search of the objective function minimum is restricted to the rectangular grid and explores a grid-restricted vicinity of the current design. Failure in making the step improve the current design leads to refining the grid size and allowing

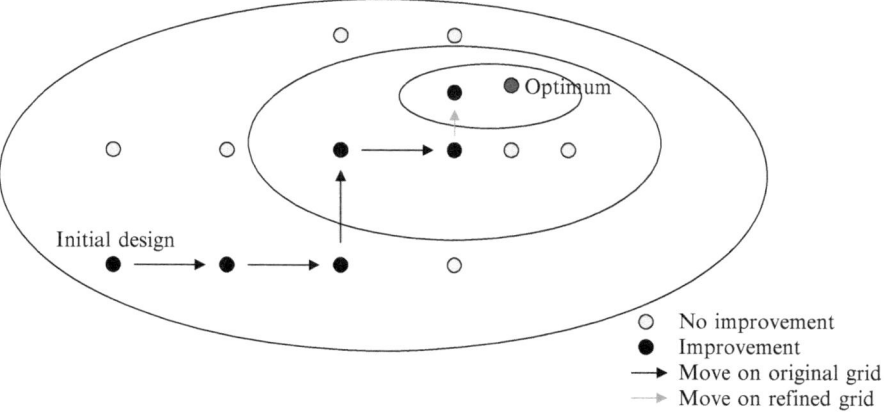

Fig. 2.3 The concept of pattern search. The search is based on exploratory movements restricted to the rectangular grid around the initial design. Upon failure of making the successful move, the grid is refined to allow smaller steps. The actual implementations of pattern search routines also use more sophisticated strategies (e.g., grid-restricted line search)

smaller steps. Various variants of the pattern search methods are available (see, e.g., Torczon 1997; Kolda et al. 2003). With sufficiently large size of the initial grid, these techniques can be used to perform a quasi-global search.

One of the most famous derivative-free methods is the Nelder-Mead algorithm (Yang 2010) also referred to as the simplex method. Its search process is based on moving the vertices of the simplex in the design space in such a way that the vertex corresponding to the worst (i.e., highest) value of the objective function is replaced by the new one at the location where the objective function value is expected to be improved.

Pattern search and similar methods are usually robust although their convergence is relatively slow compared to gradient-based routines. Their fundamental advantage is in the fact that they do not use derivative information and, even more importantly, they are quite immune to the numerical noise. An excellent and mathematically rigorous treatment of derivative-free optimization techniques, including model-based trust-region derivative-free methods, can be found in Conn et al. (2009). Many pattern search methods and their extensions possess mathematically rigorous convergence guarantees (Conn et al. 2009). An interesting extension of pattern search to constrained black-box optimization is Mesh Adaptive Direct Search (MADS) (Audet and Dennis 2006).

2.4 Metaheuristics and Global Optimization

Metaheuristics are global search methods that are based on observation of natural processes (e.g., biological or social systems). Most metaheuristics process sets (or populations) of potential solutions to the optimization problem at hand in a way that these solutions (also called individuals) interact with each other so that the optimization process is capable to avoid getting stuck in local optima and converge—with reasonable probability—to a globally optimal solution of the problem. At the same time, metaheuristics can handle noisy, non-differentiable, and discontinuous objective functions.

The most popular types of metaheuristic algorithms include genetic algorithms (GAs) (Goldberg 1989), evolutionary algorithms (EAs) (Back et al. 2000), evolution strategies (ES) (Back et al. 2000), particle swarm optimizers (PSO) (Kennedy et al. 2001), differential evolution (DE) (Storn and Price 1997), and, more recently, firefly algorithm (Yang 2010). A famous example of metaheuristic algorithm that processes a single solution rather than a population of individuals is simulated annealing (Kirkpatrick et al. 1983).

The typical flow of the metaheuristic algorithm is the following (here, P is the set of potential solutions to the problem at hand, also referred to as a population):

1. Initialize population P.
2. Evaluate population P.
3. Select parent individuals S from P.

4. Apply recombination operators to create a new population P from parent individuals S.
5. Apply mutation operators to introduce local perturbations in individuals of P.
6. If termination condition is not satisfied, go to 2.
7. End.

Initialization of the population is usually random. In the next stage, each individual is evaluated, and its corresponding value of the objective function determines its "fitness." An important step is selection of the subset of individuals to form a new population. Depending on the algorithm, the selection can be deterministic (pick up the best ones only, ES) or partially random (probability of being selected depends on the fitness value but there is a chance even for poor individuals, EAs). In some algorithms, such as PSO or DE, there is no selection at all (i.e., individuals are modified from iteration to iteration but never die). There are two types of operations that are used to modify individuals: exploratory ones (e.g., crossover in EAs or ES) and exploitative ones (e.g., mutation in GAs). Exploratory operators combine information contained in the parent individuals to create a new one. For example, in case of an evolutionary algorithm with natural (floating point) representation, a new individual c can be created as a convex combination of the parents p_1 and p_2, i.e., $c = \alpha p_1 + (1 - \alpha)p_2$, where $0 < \alpha < 1$ is a random number. These types of operators allow making large "steps" in the design space and, therefore, explore new and promising regions. Exploitative operators introduce small perturbations (e.g., $p \leftarrow p + r$, where r is a random vector selected according to a normal probability distribution with zero mean and certain, problem-dependent variance). These operators allow exploitation of a given region of the design space improving local search properties of the algorithm. In some of the more recent algorithms, e.g., PSO, the difference between both types of operators is not that clear (modification of the individual may be based on the best solution found so far by that given individual as well as the best solution found by the entire population).

A common feature of metaheuristics is that they normally require a large number of objective function evaluations to converge. Typical population size is anything between 10 and 100, whereas the number of iteration may be a few dozen to a few hundred. Also, their performance may be heavily dependent on the values of control parameters, which may not be easy to determine beforehand. Finally, because they are stochastic methods, a solution obtained in each run of the algorithm will be generally different from the previous one. From the point of view of antenna design, metaheuristics are attractive approaches for problems where evaluation time of the objective function is not of a primary concern and when multiple local optima are expected. For that reason, metaheuristics are mostly used for solving antenna array optimization problems with non-coupled radiators, in particular, for pattern synthesis (e.g., Ares-Pena et al. 1999; Bevelacqua and Balanis 2007; Li et al. 2008). It should also be mentioned that the problem of high computational cost can be partially alleviated by surrogate-assisted metaheuristics (e.g., Ong et al. 2003; Emmerich et al. 2006; Zhou et al. 2007; Jin 2011; Parno et al. 2012; Loshchilov et al. 2012; Regis 2013a, b), where metaheuristic optimization is combined with response surface modeling of the expensive objective function.

2.5 Challenges of Conventional Optimization Toward Design Using Surrogate Models

The optimization methods considered in this chapter attempt to solve the design problem (2.1) directly. In this sense, we refer to these techniques as conventional ones. As explained in Sect. 2.1 and Fig. 2.2, the direct approach requires that each new candidate design produced by the optimizer is verified by performing EM simulation of the underlying antenna structure. Because each high-fidelity simulation is already computationally expensive, conventional optimization with its large number of objective function evaluations may be prohibitive. Numerical noise that is inherent to EM simulations poses additional problems, particularly for gradient-based methods. Consequently, application of conventional off-the-shelf optimization algorithms for EM-based antenna design often results in failures. As a result, although almost all commercial simulation tools offer certain built-in optimization capabilities (e.g., CST 2013; FEKO 2012), many designers rely on repetitive parameter sweeps and own expert knowledge that allow them to find at least satisfactory designs in reasonable time.

The surrogate-based optimization concept that is a main topic of this book attempts to address this problem by replacing direct optimization of the high-fidelity model, with optimization of its cheap and analytically tractable representation referred to as a surrogate model. As indicated in the following chapters, it is possible to construct and exploit such representations in such a way that—with occasional reference to the high-fidelity model—a satisfactory design can be found at a fraction of a computational effort required by conventional optimization.

Chapter 3
Surrogate-Based Optimization

In this chapter, the surrogate-based optimization (SBO) paradigm is formulated. We discuss SBO on a generic level, including the optimization flow, fundamental properties of the SBO process, and typical ways of constructing the surrogate. We emphasize a distinction between function approximation and physics-based surrogates as well as discuss the issues of exploration and exploitation of the design space in the context of SBO.

3.1 Surrogate-Based Optimization Basics

Difficulties of the conventional optimization techniques in the context of simulation-driven design are the main motivation for developing alternative design methods. As mentioned in Chaps. 1 and 2, the major bottleneck of direct optimization of EM-based antenna models is the computational cost of accurate, high-fidelity simulation. Surrogate-based optimization (SBO) (Queipo et al. 2005; Koziel et al. 2011c; Koziel and Ogurtsov 2011b) seems to be a right approach to address this problem.

The key idea of SBO is that direct optimization of an expensive model (here, accurate high-fidelity EM simulation of the antenna structure at hand) is replaced by an iterative process, where the prediction about the optimum design comes from optimizing a fast representation of the high-fidelity model, referred to as a surrogate. Using the high-fidelity model evaluation at this predicted optimum (and, perhaps some other designs), the surrogate is enhanced and used again to find a better design (Koziel et al. 2011c). In more rigorous terms, the SBO design process can be represented as (Koziel et al. 2011c, Fig. 3.1)

$$x^{(i+1)} = \arg\min_{x} R_s^{(i)}(x) \tag{3.1}$$

where $x^{(i)}$, $i = 0, 1, \ldots$, is a sequence of approximated solutions to the original problem (2.1), whereas $R_s^{(i)}$ is the surrogate model at iteration i; $x^{(0)}$ is an initial design. The surrogate model is assumed to be computationally much cheaper than the

S. Koziel and S. Ogurtsov, *Antenna Design by Simulation-Driven Optimization*,
SpringerBriefs in Optimization, DOI 10.1007/978-3-319-04367-8_3,
© Slawomir Koziel and Stanislav Ogurtsov 2014

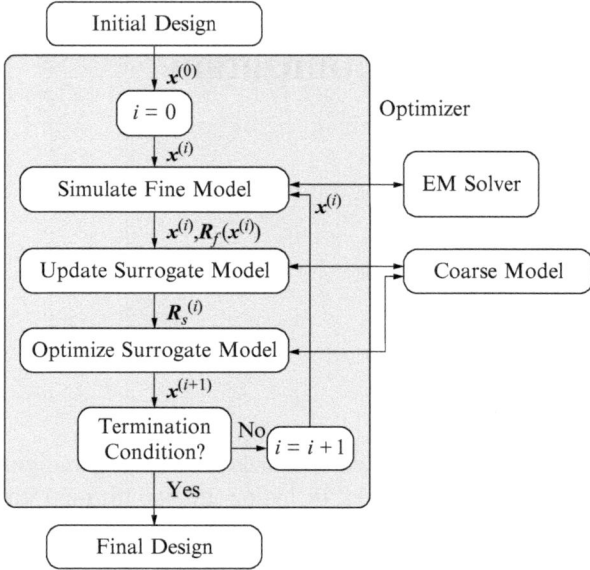

Fig. 3.1 Surrogate-based optimization concept: an approximate high-fidelity model minimizer is obtained iteratively by optimizing the surrogate model. The high-fidelity model is evaluated at each new design for verification purposes. If the termination condition is not satisfied, the surrogate model is updated and the search continues. In most cases the high-fidelity model is evaluated only once per iteration. The number of iterations needed in SBO is often substantially smaller than those in conventional (direct) optimization techniques

high-fidelity model R_f. Also, it is assumed to be sufficiently accurate, particularly in the vicinity of the current design $x^{(i)}$. If this is the case, the algorithm (3.1) is likely to quickly approach the high-fidelity optimum x^*. Typically, the high-fidelity model is evaluated only once per iteration (at every new design $x^{(i+1)}$). These data are used for design verification but also to update the surrogate model.

Because the surrogate model is computationally cheap, the optimization cost associated with (3.1) can—in many cases—be viewed as negligible, so that the total optimization cost is determined by the evaluation of the high-fidelity model. Normally, the number of iterations needed by the SBO algorithm is substantially smaller than that needed by majority of methods that optimize the high-fidelity model directly (e.g., gradient-based schemes with numerical derivatives) (Koziel et al. 2006).

Some of the SBO methods, particularly space mapping (Bandler et al. 2004a, b; Koziel et al. 2008b), that have been recently applied to antenna design originated in microwave engineering where circuit equivalents or even analytical formulas are commonly used to construct the surrogate model (Bandler et al. 2004a, b). In such cases, the evaluation time of the surrogate can indeed be neglected. Unfortunately, in the case of antenna design, this assumption is rarely satisfied. This is because reliable antenna surrogates based on equivalent circuits are hardly available. The most common and universal way of creating surrogate models for antennas is through coarse-discretization EM simulations (see Chap. 5 for details). Therefore the computational cost of multiple evaluations of the antenna surrogate while solving (3.1) is not negligible. This poses an additional challenge for designing SBO algorithms for antenna design to ensure that not only the number of high-fidelity but also surrogate model evaluations is small as possible.

Formally speaking, the SBO algorithm (3.1) is provably convergent to a local optimum of R_f (Alexandrov et al. 1998) if the surrogate model satisfies zero- and first-order consistency conditions with the high-fidelity model (i.e., $R_s^{(i)}(x^{(i)})=R_f(x^{(i)})$ and $J[R_s^{(i)}(x^{(i)})]=J[R_f(x^{(i)})]$ (Alexandrov and Lewis 2001)), where $J[\cdot]$ stands for the model Jacobian, and the surrogate-based algorithm is enhanced by a trust-region mechanism (Conn et al. 2000):

$$x^{(i+1)} = \arg\min_x U(R_s^{(i)}(x)) \quad subject\ to \quad \| x - x^{(i)} \| \le \delta^{(i)} \tag{3.2}$$

where $\delta^{(i)}$ denotes the trust-region radius at iteration i. The trust region is updated at every iteration (Conn et al. 2000). The idea of the trust-region approach is that if the vicinity of $x^{(i)}$ is sufficiently small, the first-order consistent surrogate becomes—in this vicinity—a sufficiently good representation of R_f to produce a design that reduces the high-fidelity objective function value upon completing (3.2).

It should be stressed that in order to satisfy the first-order consistency condition, both high-fidelity and surrogate model sensitivity data are required. Formally speaking, some additional assumptions concerning the smoothness of the functions involved are also necessary for convergence (Echeverría and Hemker 2008). Convergence of the SBO algorithm can also be guaranteed under various other settings; see e.g., Koziel et al. (2006) and Koziel et al. (2008a) (space mapping), Echevería and Hemker (2008) (manifold mapping), or Booker et al. (1999) (surrogate management framework).

The SBO scheme (3.1) can be, in general, implemented as a local or global optimization. In practice, particularly if the trust-region-like convergence safeguards (3.2) are used and the surrogate model is first-order consistent, the SBO scheme is typically executed as a local search. If the surrogate model is constructed globally over the entire design space, it is possible to turn the process (3.2) into a global optimization algorithm. In this case the surrogate model can be optimized using, for example, evolutionary algorithms and updated using certain statistical infill criteria based on the expected improvement of the objective function or minimization of the (global) modeling error (see also Sect. 3.3).

3.2   Surrogate Model Construction: Function Approximation and Physics-Based Surrogates

Over the last years, a number of surrogate-based optimization techniques have been proposed (Alexandrov and Lewis 2001; Queipo et al. 2005; Bandler et al. 2004a, b; Koziel 2010a). Some of these techniques are quite recent and they have been developed specifically to solve problems in microwave engineering, e.g., space mapping (Bandler et al. 2004a, b) or shape-preserving response prediction (Koziel 2010a).

Fig. 3.2 Approximation-
based surrogate modeling
flow. If quality of the model
is not sufficient, the
procedure should be iterated
(additional data points should
be acquired)

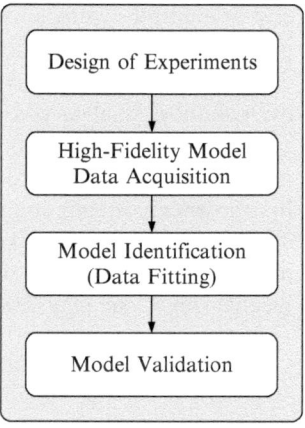

Most of the SBO methods are based on the same principle described in Sect. 3.1.
The differences between these techniques are not only in the way they operate but,
more importantly, in the way of constructing the surrogate model $s(x)$ of the high-
fidelity model $f(x)$. The latter is a key component of any SBO algorithm. It has to be
computationally cheap, preferably smooth, and, at the same time, reasonably accurate, so that it can be used to predict an approximate location of the high-fidelity
model minimizer.

Broadly speaking, there are two major types of surrogate models: approximation-
and physics-based ones. Approximation (or functional) surrogate models are constructed through approximations of the high-fidelity model data obtained by
sampling the design space using appropriate design of experiments (DOE) methodologies (Queipo et al. 2005). Modeling flows for approximation-based surrogates,
strategies for allocating samples (Simpson et al. 2001), generating approximations
(Queipo et al. 2005; Forrester and Keane 2009; Simpson et al. 2001), as well as
validating the surrogates are discussed in Sect. 3.2.1. Another modeling approach is
to exploit some knowledge about the system under consideration embedded in the
physics-based low-fidelity model. In case of antennas, the low-fidelity model is
typically obtained from coarse-discretization EM simulations. Then the surrogate is
constructed out of the low-fidelity model by applying appropriate correction
techniques. The concept and discussion of physics-based surrogates is presented in
Sect. 3.2.2. The surrogate-based optimization techniques exploiting physics-based
models—the main topic of this book—are described in Chap. 4.

3.2.1 Approximation-Based Surrogate Models

Approximation-based models are probably the most popular class of surrogates due
to a large variety of techniques that are available in the literature and implemented
as ready-to-use toolboxes, particularly in Matlab (Matlab 2012).

Approximation-based models are constructed as approximations to high-fidelity
model data. The first step of the modeling process shown in Fig. 3.2 is allocation of

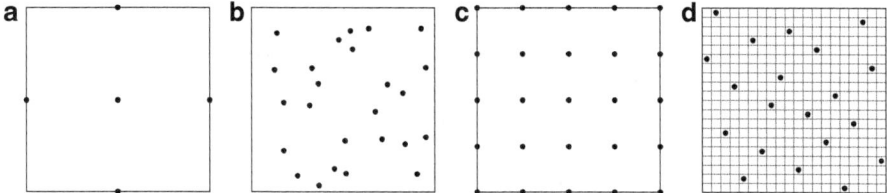

Fig. 3.3 Popular DOE techniques: (**a**) one of possible factorial designs (star distribution); (**b**) random sampling; (**c**) uniform grid sampling; (**d**) Latin hypercube sampling (LHS)

training samples in the design space. This step is referred to as design of experiments (DOE) (Giunta et al. 2003; Santner et al. 2003; Koehler and Owen 1996). Figure 3.3 shows a few popular DOE strategies, including factorial design, random sampling, uniform grid sampling, and Latin hypercube sampling (LHS) (Beachkofski and Grandhi 2002) which is perhaps the most popular uniform sampling method used today.

Factorial design (allocating samples in the corners, edges, and/or faces of the design space) (Fig. 3.3a) is a traditional DOE approach that allows estimating the main effect and interactions between design variables without having too many samples. Nowadays, uniform sampling methods are more popular. Random sampling shown in Fig. 3.3b allows allocating any number of samples; however, sample uniformity is rather poor. The best uniformity can be obtained using grid sampling (Fig. 3.3c), but this method does not permit an arbitrary number of samples. LHS overcomes the drawbacks of random sampling by improving uniformity (Giunta et al. 2003).

Having allocated the data samples $\{x^{(i)}\}$, $i = 1, \ldots, p$, training data is acquired by evaluating the high-fidelity model (in case of antennas, it is usually full-wave EM analysis). The data pairs $\{x^{(i)}, f(x^{(i)})\}$ are then approximated. A number of approximation methods are available, including polynomial regression (Queipo et al. 2005), radial basis functions (Wild et al. 2008), kriging (Forrester and Keane 2009), neural networks (Haykin 1998), support vector regression (Gunn 1998), Gaussian process regression (Rasmussen and Williams 2006), and rational approximation (Shaker et al. 2009) to name a few. For example, the linear regression model usually takes the form of

$$s(x) = \sum_{j=1}^{K} \beta_j v_j(x) \tag{3.3}$$

where β_j are unknown coefficients and v_j are basis functions. The model parameters can be found as a least-squares solution to the linear system

$$f = X\beta \tag{3.4}$$

where $f = [f(x^{(1)})\ f(x^{(2)}) \ldots f(x^{(p)})]^T$, X is a $p \times K$ matrix containing the basis functions evaluated at the sample points, and $\beta = [\beta_1\ \beta_2 \ldots \beta_K]^T$. One of the simplest examples of a regression model is a second-order polynomial one defined as

$$s(x) = s\left(\left[x_1 x_2 \dots x_n\right]^{\mathrm{T}}\right) = \beta_0 + \sum_{j=1}^{n} \beta_j x_j + \sum_{i=1}^{n}\sum_{j\leq i}^{n} \beta_{ij} x_i x_j \qquad (3.5)$$

with the basis functions being monomials: 1, x_j, and $x_i x_j$.

A special case of a linear regression model is a radial basis function model,

$$s(x) = \sum_{j=1}^{K} \lambda_j \phi\left(\left\|x - c^{(j)}\right\|\right), \qquad (3.6)$$

where $\lambda = [\lambda_1 \lambda_2 \dots \lambda_K]^{\mathrm{T}}$ is the vector of model parameters and $c^{(j)}$, $j = 1, \dots, K$, are the (known) basis function centers. A popular choice of the basis function is a Gaussian, $\phi(r) = \exp(-r^2/2\sigma^2)$, where σ is the scaling parameter.

A popular technique to interpolate deterministic noise-free data is kriging (Journel and Huijbregts 1981; Simpson et al. 2001; Kleijnen 2009; O'Hagan 1978). Kriging is a Gaussian process-based modeling method, which is compact and cheap to evaluate (Rasmussen and Williams 2006). In its basic formulation, kriging (Journel and Huijbregts 1981; Simpson et al. 2001) assumes that the function of interest is of the following form:

$$f(x) = g(x)^{\mathrm{T}} \beta + Z(x), \qquad (3.7)$$

where $g(x) = [g_1(x) \ g_2(x) \dots g_K(x)]^{\mathrm{T}}$ are known (e.g., constant) functions, $\beta = [\beta_1 \ \beta_2 \dots \beta_K]^{\mathrm{T}}$ are the unknown model parameters, and $Z(x)$ is a realization of a normally distributed Gaussian random process with zero mean and variance σ^2. The regression part $g(x)^{\mathrm{T}}\beta$ globally approximates the function f, and $Z(x)$ takes into account localized variations. The covariance matrix of $Z(x)$ is given as

$$Cov\left[Z\left(x^{(i)}\right)Z\left(x^{(j)}\right)\right] = \sigma^2 R\left(\left[R\left(x^{(i)}, x^{(j)}\right)\right]\right) \qquad (3.8)$$

where R is a $p \times p$ correlation matrix with $R_{ij} = R(x^{(i)}, x^{(j)})$. Here, $R(x^{(i)}, x^{(j)})$ is the correlation function between sampled data points $x^{(i)}$ and $x^{(j)}$. The most popular choice is the Gaussian correlation function

$$R(x, y) = \exp\left[-\sum_{k=1}^{n} \theta_k \mid x_k - y_k \mid^2\right] \qquad (3.9)$$

where θ_k are unknown correlation parameters and x_k and y_k are the kth component of the vectors x and y, respectively. The kriging predictor (Simpson et al. 2001; Journel and Huijbregts 1981) is defined as

$$s(x) = g(x)^{\mathrm{T}} \beta + r^{\mathrm{T}}(x) R^{-1}(f - G\beta) \qquad (3.10)$$

where $r(x) = [R(x, x^{(1)}) \dots R(x, x^{(p)})]^{\mathrm{T}}$, $f = [f(x^{(1)}) f(x^{(2)}) \dots f(x^{(p)})]^{\mathrm{T}}$, and G is a $p \times K$ matrix with $G_{ij} = g_j(x^{(i)})$. The vector of model parameters β can be computed as $\beta = (G^{\mathrm{T}}R^{-1}G)^{-1}G^{\mathrm{T}}R^{-1}f$. Model fitting is accomplished by maximum likelihood for θ_k (Journel and Huijbregts 1981).

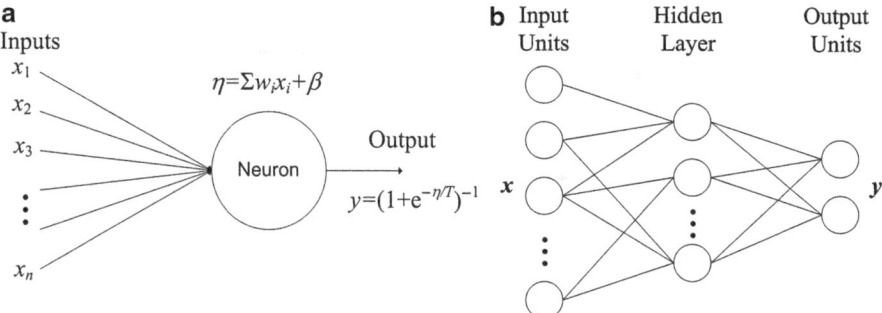

Fig. 3.4 Neural networks: (**a**) neuron basic structure; (**b**) two-layer feedforward neural network architecture

Perhaps the most popular type of the approximation-like modeling approach in microwave engineering is artificial neural networks (ANNs) (Rayas-Sánchez 2004; Kabir et al. 2008). The basic structure in a neural network (Haykin 1998; Minsky and Papert 1969) is the neuron (or single-unit perceptron). A neuron performs an affine transformation followed by a nonlinear operation (see Fig. 3.4a). If the inputs to a neuron are denoted as $x_1, \ldots, x_n$, the neuron output y is computed as

$$y = \frac{1}{1 + \exp\left(-\eta / T\right)} \tag{3.11}$$

where $\eta = w_1 x_1 + \cdots + w_n x_n + \gamma$, with $w_1, \ldots, w_n$ being regression coefficients. Here, γ is the bias value of the neuron, and T is a user-defined (slope) parameter. Neurons can be combined in multiple ways (Haykin 1998). The most common neural network architecture is the multilayer feedforward network (see Fig. 3.4b).

The construction of a neural network model requires two main steps: (1) architecture selection and (2) network training. The network training can be stated as a nonlinear least-squares regression problem for a number of training points. A popular technique for solving this regression problem is the error back-propagation algorithm (Simpson et al. 2001; Haykin 1998).

In identifying the surrogate model parameters, the objective is usually to minimize the training error at the sample set $\{x^{(i)}\}$. In some cases, the surrogate model parameters can be calculated analytically (e.g., with polynomial regression); in others, they are obtained by solving a separate minimization problem (e.g., kriging, neural networks, etc.). The training error can be defined using a norm, e.g., $\|s(x^{(i)}) - f(x^{(i)})\|$ averaged over all samples. The surrogate model identification should be done in such a way that a generalization error (i.e., the error at the designs other than the training ones) is also as small as possible. Estimating the generalization error is referred to as model validation. In many cases, validation of the model is carried out using a separate set of testing samples (so-called split-sample method; Simpson et al. 2001). A better and one of the most popular validation methods is the so-called cross-validation (Simpson et al. 2001), where the same set of samples is used for both training and testing in the following way: a subset (say, $[(K-1)/K] \cdot p$) of available samples is used to train the model and the remaining part is used for testing. Then, the training-testing iteration is repeated with another p/K of samples used as a testing set. After repeating the procedure K times, the estimated generalization error is obtained as an average of the K estimates obtained in all iterations.

In practice, the entire procedure of allocating samples, acquiring data, model identification, and validation can be repeated a number of times until the prescribed surrogate accuracy is reached. In each repetition, a new set of training samples is added to the existing ones. The strategies of allocating new samples (so-called infill criteria; Forrester and Keane 2009) usually aim at improving the global accuracy of the model, i.e., inserting new samples at the locations where the estimated modeling error is the highest.

From the antenna optimization standpoint, the main advantage of approximation surrogates is that they are very fast. Unfortunately, the high computational cost of setting up such models (mostly due to acquiring the training data) is a significant disadvantage. In order to ensure decent accuracy, hundreds or even thousands of data samples are required, and the number of training points quickly grows with dimensionality of the design space. Therefore, approximation surrogates are mostly suitable to build multiple-use library models. Their use for ad hoc antenna optimization is rather limited.

3.2.2 Physics-Based Surrogate Models

A physics-based surrogate is created by correcting (or enhancing) the underlying low-fidelity model that is a simplified representation of the structure under design. In microwave engineering, a popular choice of the low-fidelity model is a circuit equivalent because it is fast and easily available for many structures, e.g., filters (Bandler et al. 2004a, b). In case of antennas, the only universally available way of obtaining low-fidelity models is through coarse-discretization EM simulation. A discussion of low-fidelity antenna models is presented in Chap. 5 of this book.

The main advantage of physics-based models is that—because of exploiting some knowledge embedded in the low-fidelity model—a limited amount of high-fidelity data is necessary to ensure decent accuracy. For the same reason, physics-based surrogates are characterized by good generalization capability, i.e., they can provide reliable prediction of the high-fidelity model response at the designs not used in the training process. These advantages are normally translated into better efficiency (in particular, lower CPU cost) when physics-based surrogates are used in the design optimization process (Koziel et al. 2011c).

The remaining part of this book is entirely focused on surrogate-based optimization methods exploiting physics-based models. Chapter 4 provides details regarding a number of such methods, where—in some cases—construction of the surrogate model is an inherent part of the optimization process. In this section, we present a few elementary ways of correcting the low-fidelity model in order to construct the physics-based surrogate.

Let $c(x)$ denote a low-fidelity model of the device of interest. One of the simplest ways of constructing a surrogate of a high-fidelity model f is a response correction. In the context of the iterative design optimization (3.1) that produces a sequence $\{x^{(i)}\}$ of approximate solutions to the original problem (2.1), particularly if the algorithm is embedded in the trust-region framework (cf. (3.2)), a local alignment between the surrogate and the high-fidelity model is of fundamental importance. Then, the surrogate $s^{(i)}(x)$ at iteration i can be constructed as

$$s^{(i)}(x) = \beta_i(x)c(x), \tag{3.12}$$

where $\beta_i(x) = \beta_i(x^{(i)}) + \nabla\beta(x^{(i)})^{\mathrm{T}}(x - x^{(i)})$ and where $\beta(x) = f(x)/c(x)$. This ensures satisfaction of the so-called zero- and first-order consistency conditions between s and f, i.e., agreement of function values and their gradients at $x^{(i)}$ (Alexandrov and Lewis 2001).

Another way of correcting the low-fidelity model is so-called input space mapping (ISM) (Bandler et al. 2004a, b), where the surrogate is created as (here, we use vector notation for the models, i.e., R_c and R_s for the low-fidelity and the surrogate models, respectively)

$$R_s^{(i)}(x) = R_c\left(x + c^{(i)}\right) \tag{3.13}$$

with the model parameters $c^{(i)}$ obtained by minimizing $\|R_f(x^{(i)}) - R_c(x^{(i)} + c^{(i)})\|$. Figure 3.5 shows an example of a filter structure evaluated using EM simulation (high-fidelity model), its circuit equivalent (low-fidelity model), and the corresponding $|S_{21}|$ responses before and after applying the ISM correction.

In many cases, the major type of discrepancy between the low- and high-fidelity models is a frequency shift. In these situations, misalignment between the models can be reduced by using frequency scaling or frequency space mapping (Koziel et al. 2006). We assume that $R_f(x) = [R_f(x,f_1)\ R_f(x,f_2)\ \dots\ R_f(x,f_m)]^{\mathrm{T}}$, where $R_f(x,f_k)$ is the evaluation of the high-fidelity model at a frequency f_k, whereas f_1 through f_m represent the entire discrete set of frequencies at which the model is evaluated. Similar convention is used for the low-fidelity model. The frequency-scaled model $S_{c\cdot F}(x)$ is defined as

$$R_s^{(i)}(x) = \left[R_c\left(x, F_0 + F_1 f_1\right)\ \dots\ R_c\left(x, F_0 + F_1 f_m\right)\right]^{\mathrm{T}}, \tag{3.14}$$

where F_0 and F_1 are scaling parameters obtained to minimize misalignment between R_s and R_f at $x^{(i)}$ as

$$[F_0, F_1] = \arg\min_{[F_0, F_1]} R_f\left(x^{(i)}\right) - R_s\left(x^{(i)}\right). \tag{3.15}$$

Figure 3.6 shows an example of frequency scaling applied to the low-fidelity model of a substrate-integrated cavity antenna. Here, both the low- and high-fidelity models are evaluated using EM simulation (R_c with coarse discretization).

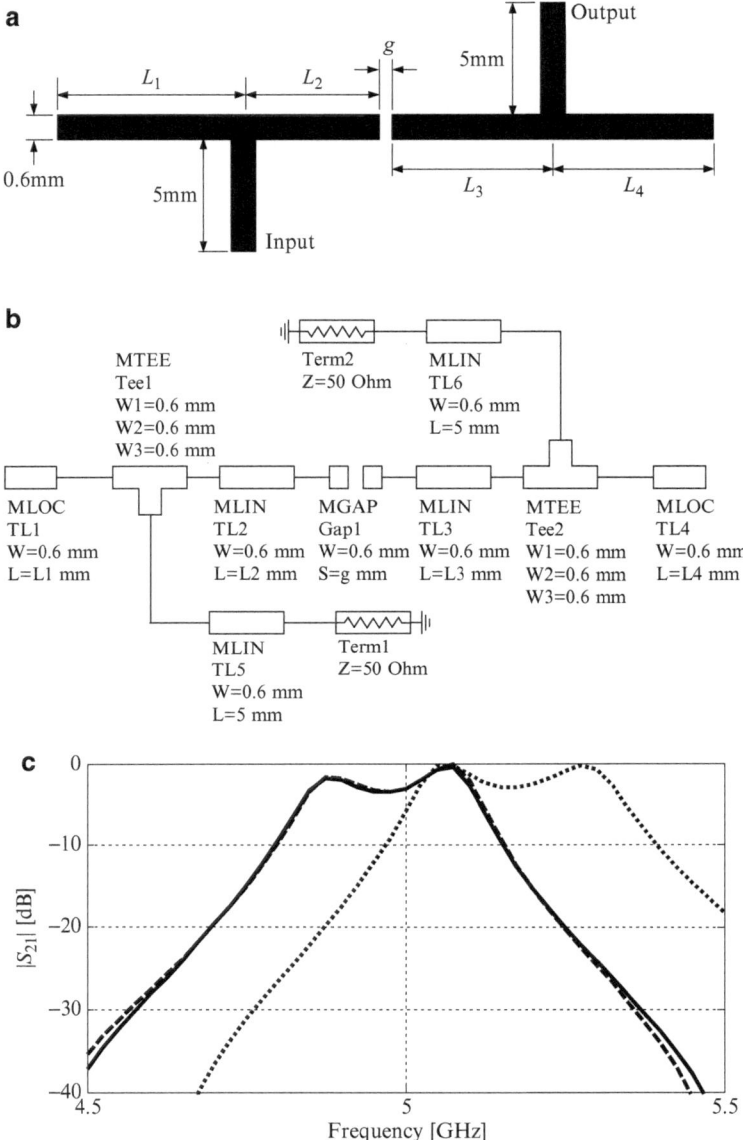

Fig. 3.5 Low-fidelity model correction through parameter shift (input space mapping): (a) microstrip filter geometry (high-fidelity model R_f evaluated using EM simulation); (b) low-fidelity model R_c (equivalent circuit); (c) response of R_f (*solid line*) and R_c (*dotted line*), as well as response of the surrogate model R_s (*dashed line*) created using input space mapping as in (3.13)

More information about physics-based surrogates can be found in Chap. 4. It should be emphasized that simple correction schemes such as those described here are often used as building blocks to construct more involved surrogate models (Koziel et al. 2006).

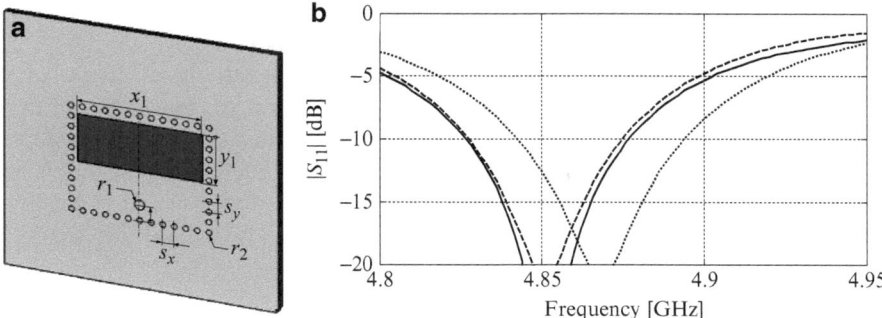

Fig. 3.6 Low-fidelity model correction through frequency scaling: (**a**) antenna geometry (both R_f and R_c evaluated using EM simulation, coarse discretization used for R_c); (**b**) response of R_f (*solid line*) and R_c (*dotted line*), as well as response of the surrogate model R_s (*dashed line*) created using frequency scaling as in (3.14) and (3.15)

3.3 Exploration Versus Exploitation

The surrogate-based optimization process starts from an initial surrogate model which is updated using the high-fidelity model data that is accumulated in the optimization process. In particular, the high-fidelity model has to be evaluated for verification at any new design $x^{(i)}$ provided by the surrogate model. The new designs at which we evaluate the high-fidelity model are referred to as infill points (Forrester and Keane 2009) and selection of the infill points is also known as adaptive sampling (Forrester and Keane 2009). For most SBO methods discussed in this book, infill points are selected through local optimization of the surrogate. This is usually justified because the initial design is assumed to be reasonably good (in practice, it is obtained through local or global optimization of the low-fidelity model). This choice of infill criteria aims at the exploitation of a certain region of the design space, more specifically, vicinity of a local optimum that is close to the initial design.

The exploration of the design space implies in most cases a global search. If the underlying objective function is non-convex, exploration usually boils down to performing a global sampling of the search space, for example, by selecting those points that maximize some estimation of the error associated to the surrogate considered (Forrester and Keane 2009). It should be stressed that global exploration is often impractical, especially for computationally expensive cost functions with a medium/large number of optimization variables. Additionally, pure exploration may not be a good approach for updating the surrogate from the optimization point of view since a great amount of computing resources might be spent in modeling parts of the search space that are either infeasible or far to optimality.

Therefore, it appears that in optimization there should be a balance between exploitation and exploration. As suggested by Forrester and Keane (2009), this trade-off could be formulated in the context of surrogate-based optimization, for example, by means of a bi-objective optimization problem with a global measure of the error associated to the surrogate as a second objective function.

Surrogate-based methods that utilize infill criteria as a part of their formulation, i.e., balancing the need for exploration versus exploitation, are often referred to as efficient global optimization (EGO) techniques. A detailed discussion of EGO is given by Jones et al. (1998) and by Forrester and Keane (2009). Similar approaches have also been described in Villemonteix et al. (2009), Björkman and Holmström (2000), as well as Gutmann (2001). The last two exploit radial basis function surrogates rather than kriging.

Chapter 4
Methodologies for Variable-Fidelity Optimization of Antenna Structures

In this chapter, we formulate and discuss several surrogate-based optimization techniques and algorithms that may be useful for computationally efficient antenna optimization. All methods presented here exploit variable-fidelity EM simulations. In particular, in order to optimize the high-fidelity EM model R_f of the antenna structure under consideration, an auxiliary low-fidelity model R_c is utilized that is normally based on coarse-discretization EM evaluation of the same structure. Under suitable correction, the low-fidelity model provides reliable predictions regarding the improved design of the high-fidelity one. The methods of setting up the low-fidelity model are elaborated in Chap. 5. A discussion of optimization techniques is preceded, in Sect. 4.1, by outlining challenges of surrogate-based antenna design. Sections 4.2–4.7 contain formulation of specific methodologies. Applications for the design optimization of various antenna structures are covered in Chaps. 6–11.

4.1 Antenna-Specific Challenges of Surrogate-Based Optimization

A discussion of surrogate-based optimization paradigm presented in Chap. 3 indicated that the major advantage of SBO methods is in possible reduction of the computational cost of the design process they offer. One of the most important prerequisites is that the surrogate, and, consequently (as we are focused on physics-based surrogates), the underlying low-fidelity model, is substantially faster than the high-fidelity model to be optimized. Unfortunately, this requirement is hardly possible to satisfy in for antennas. In most cases, particularly for broadband and ultra-wideband antennas, dielectric resonator antennas, and substrate-integrated waveguide/cavity antennas (Balanis 2005; Petosa 2007; Volakis 2007), reliable circuit equivalents or analytical models are not available. Therefore, the only way of obtaining usable low-fidelity models for most antenna structures is through

S. Koziel and S. Ogurtsov, *Antenna Design by Simulation-Driven Optimization*,
SpringerBriefs in Optimization, DOI 10.1007/978-3-319-04367-8_4,
© Slawomir Koziel and Stanislav Ogurtsov 2014

coarse-discretization EM simulations. The problem of setting up coarse-discretization EM models for surrogate-based antenna design is addressed in detail in Chap. 5 of this book. When properly set up, coarse-discretization low-fidelity models are sufficiently accurate to be used by SBO-based design procedures. However, they are also relatively expensive, typically, only 10–50 times faster than respective high-fidelity models. This creates specific challenges while developing SBO techniques for antenna design. In particular, the total evaluation cost of the low-fidelity model in the SBO process (both due to updating and optimization of the surrogate model) cannot be neglected and can significantly contribute to the overall design cost. For that reason, SBO algorithms for antennas should aim at reducing not only the number of high- but also low-fidelity model evaluations. As we will see, this can be addressed in several ways, e.g., by creating an auxiliary response surface model (Sect. 4.2), by using more efficient surrogate modeling methods (e.g., Sect. 4.3), and by utilizing several low-fidelity models (Sect. 4.7).

4.2 Space Mapping

Space mapping (SM) (Bandler et al. 1994, 2004a, b; Bakr et al. 1999; Koziel, et al. 2008b) is one of the earliest and most popular surrogate-based optimization techniques in microwave engineering. Here, we discuss the original SM concept, aggressive SM, parametric SM, as well as SM with auxiliary response surface approximation model, which is suitable for antenna design.

4.2.1 Space Mapping Concept

The original SM idea was based on a concept of mapping P relating the high- and low-fidelity model parameters (Bandler et al. 1994)

$$x_c = P\left(x_f\right) \tag{4.1}$$

so that $R_c(P(x_f)) \approx R_f(x_f)$ at least in some subset of the fine model parameter space. One of the issues here is to ensure one-to-one correspondence, which usually holds in practice, at least, locally (Bandler et al. 2004a, b). Having the mapping P, the direct solution of the original problem (2.1) can be replaced by finding $x_f^{\#} = P^{-1}(x_c^*)$. Here, x_c^* is the optimal design of R_c defined as $x_c^* = \mathrm{argmin}\{x_c: U(R_c(x_c))\}$, whereas $x_f^{\#}$ can be considered as a reasonable estimate of x_f^*. Using this concept, the problem (2.1) can be reformulated as

$$x_f^{\#} = \arg\min_{x_f} U\left(R_c\left(P\left(x_f\right)\right)\right) \tag{4.2}$$

where $R_c(P(x_f))$ is an enhanced low-fidelity model (or, the surrogate). Unfortunately, the space mapping P in (4.4) is not known explicitly: it can only be evaluated at any given x_f by performing the so-called parameter extraction (PE) procedure

$$P\left(x_f\right) = \arg\min_{x_c} \| R_f\left(x_f\right) - R_c\left(x_c\right)\| \tag{4.3}$$

The issues of original SM, such as nonuniqueness of the solution to (4.3) (Bandler et al. 1995) and possible misalignment of high- and low-fidelity model ranges (Alexandrov and Lewis 2001), led to numerous improvements, including parametric SM (cf. Sect. 4.2.3).

4.2.2   Aggressive Space Mapping

A popular version of SM based on the original concept is aggressive SM (ASM) (Bandler et al. 1995). If the uniqueness of the low-fidelity model optimum x_c^* is assumed, the solution to (4.2) is equivalent to reducing the residual vector $f = f(x_f) = P(x_f) - x_c^*$ to zero. The ASM technique iteratively solves the nonlinear system

$$f\left(x_f\right) = 0 \tag{4.4}$$

for x_f. The first step of the ASM algorithm is to find x_c^*. At jth iteration, the calculation of the error vector $f^{(j)}$ requires an evaluation of $P^{(j)}(x_f^{(j)})$, which is realized by executing parameter extraction (4.3), i.e., $P(x_f^{(j)}) = \arg\min\{x_c: R_f(x_f^{(j)}) - R_f(x_c)\|\}$. The quasi-Newton step in the fine model parameter space is given by

$$B^{(j)} h^{(j)} = -f^{(j)} \tag{4.5}$$

where $B^{(j)}$ is the approximation of the space mapping Jacobian $J_P = J_P(x_f) = [\partial P^T/\partial x_f]^T = [\partial(x_c^T)/\partial x_f]^T$. Solving (4.5) for $h^{(j)}$ gives the next iterate $x_f^{(j+1)}$

$$x_f^{(j+1)} = x_f^{(j)} + h^{(j)} \tag{4.6}$$

The algorithm terminates if $\|f^{(j)}\|$ is sufficiently small. The output of the algorithm is an approximation to $x_f^\# = P^{-1}(x_c^*)$. A popular way of obtaining the matrix B is through a rank one Broyden update (Broyden 1965) of the form

$$B^{(j+1)} = V^{(j)} + \frac{f^{(j+1)} - f^{(j)} - B^{(j)} h^{(j)}}{h^{(j)T} h^{(j)}} h^{(j)T} \tag{4.7}$$

Several variations of ASM have been considered in the literature, including hybrid ASM (Bakr et al. 1999) and trust-region ASM (Bakr et al. 1998).

4.2.3 Parametric Space Mapping

The mapping P described in Sect. 4.2.1 is not defined explicitly but through the parameter extraction process (4.3). In general, P can be assumed to have a certain analytical form, and the SM algorithm is defined through an iterative process (3.1).

Equation (3.13) shows a simple example of a so-called input SM surrogate that was the first parameterized version of space mapping (Bandler et al. 1994). More generally, the input SM surrogate model can take the form (Koziel et al. 2006)

$$R_s^{(i)}(x) = R_c\left(B^{(i)} \bullet x + c^{(i)}\right) \tag{4.8}$$

with $B^{(i)}$ and $c^{(i)}$ being matrices obtained in the parameter extraction process

$$\left[B^{(i)}, c^{(i)}\right] = \arg\min_{[B,c]} \sum_{k=0}^{i} w_{i\cdot k} \parallel R_f\left(x^{(k)}\right) - R_c\left(B \bullet x^{(k)} + c\right) \parallel \tag{4.9}$$

Here, $w_{i\cdot k}$ are weighting factors; a common choice of $w_{i\cdot k}$ is $w_{i\cdot k}=1$ for all i and all k (all previous designs contribute to the parameter extraction process) or $w_{i\cdot i}=1$ and $w_{i\cdot k}=0$ for $k<i$ (the surrogate model depends on the most recent design only). If B is an identity matrix, the surrogate (4.8) reduces to (3.13).

In general, the SM surrogate model is constructed as follows:

$$R_s^{(i)}(x) = \bar{R}_s\left(x, p^{(i)}\right) \tag{4.10}$$

where $\bar{R}_s$ is a generic SM surrogate model, i.e., the low-fidelity model R_c composed with suitable (usually linear) transformations, whereas

$$p^{(i)} = \arg\min_{p} \sum_{k=0}^{i} w_{i\cdot k} \parallel R_f\left(x^{(k)}\right) - \bar{R}_s\left(x^{(k)}, p\right) \parallel \tag{4.11}$$

is a vector of model parameters.

A number of SM surrogate models have been developed (see, e.g., Bandler et al. 2004a, b and Koziel et al. 2006). They can be roughly divided into four categories:

- Models based on a distortion of the low-fidelity model parameter space, in particular, the input SM of the form $\bar{R}_s(x,p) = \bar{R}_s(x,B,C) = R_c(B \bullet x + c)$ (Bandler et al. 1994, 2004a, b).
- Models based on a correction of the low-fidelity model response, e.g., the output SM of the form $\bar{R}_s(x,p) = \bar{R}_s(x,d) = R_c(x) + d$ or $\bar{R}_s(x,p) = \bar{R}_s(x,A) = A \cdot R_c(x)$ (Bandler et al. 2003; Koziel et al. 2006).
- Models utilizing an adjustment of an additional set of parameters, which are separate from the design variables, i.e., implicit space mapping of the form $\bar{R}_s(x,p) = \bar{R}_s(x,x_p) = R_{c\cdot i}(x,x_p)$, with $R_{c\cdot i}$ being the low-fidelity model dependent on both the design variables x and so-called preassigned parameters x_p (e.g., dielectric constant, substrate height) that are normally fixed in the

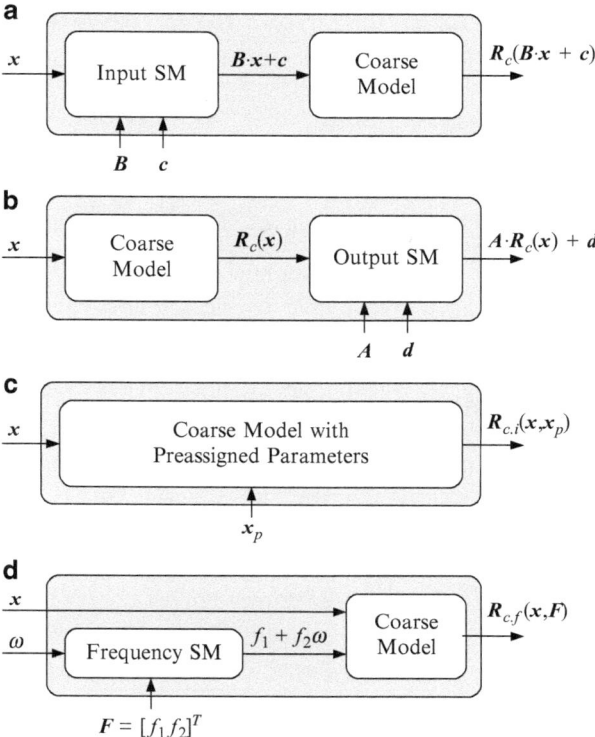

Fig. 4.1 Basic space mapping surrogate model types: (**a**) input SM, (**b**) output SM, (**c**) implicit SM, (**d**) frequency SM

high-fidelity model but can be freely altered in the low-fidelity model (Bandler et al. 2003, 2004b; Koziel et al. 2010b, 2011c).

• Special models exploiting parameters specific to a given problem. A common parameter used in microwave engineering is frequency. Frequency SM exploits a surrogate model of the form $\bar{R}_s(x,p) = \bar{R}_s(x,F) = R_{c \cdot f}(x,F)$ (Bandler et al. 2003), where $R_{c \cdot f}$ is a frequency-mapped coarse model. Here, the low-fidelity model is evaluated at frequencies different from the original frequency sweep for the high-fidelity one, according to the mapping $\omega \rightarrow f_1 + f_2\omega$, with $F = [f_1\, f_2]^T$. An illustration example of frequency SM can be found in Chap. 3 (see (3.14), (3.15), and Fig. 3.6).

Figure 4.1 illustrates the basic types of SM surrogates. Elementary SM transformations described above can be combined into more involved models (Koziel et al. 2006). For example, the surrogate utilizing the input, output, and frequency SM types takes the following form: $\bar{R}_s(x,p) = \bar{R}_s(x,c,d,F) = R_{c \cdot f}(x+c,F) + d$. In general, selection of the optimal surrogate for a given problem is not a trivial task (Koziel and Bandler 2007; Koziel et al. 2008a).

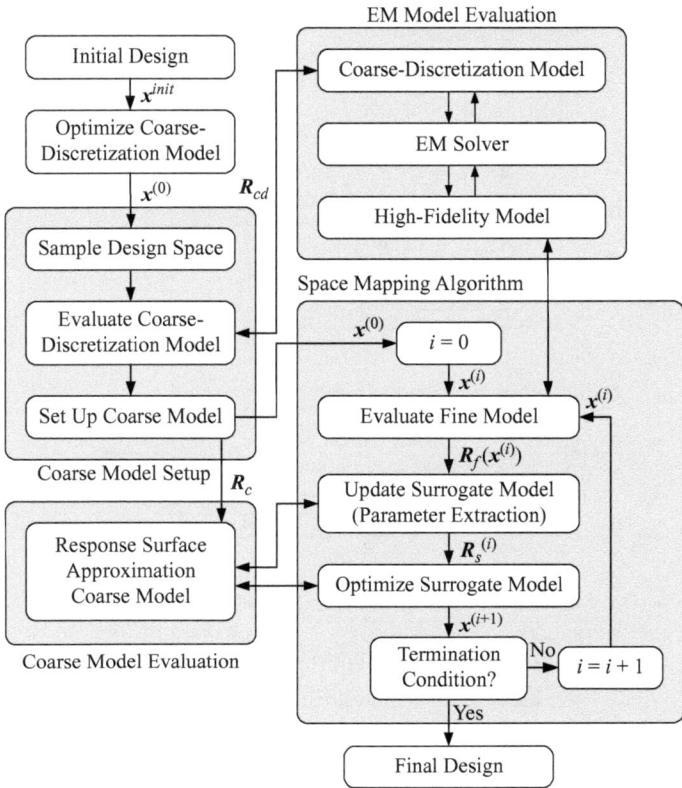

Fig. 4.2 Flowchart of the SM optimization procedure utilizing auxiliary response surface approximation coarse model (Koziel and Ogurtsov 2011d)

4.2.4 Space Mapping with Response Surface Approximations

One of the important assumptions of space mapping is that the low-fidelity model is very fast, so that the computational overhead due to surrogate model optimization (2.1) and, more importantly, parameter extraction (4.11) is negligible. Because low-fidelity antenna models are normally obtained through coarse-discretization EM simulation, overall costs of multiple evaluations of the low-fidelity model can become a bottleneck of the SBO algorithm. Therefore, it is often advantageous to construct an auxiliary response surface approximation (RSA) model of the low-fidelity one that replaces the latter in the SM optimization process (Fig. 4.2) (Koziel and Ogurtsov 2011d).

Let us denote the coarse-discretization EM low-fidelity model as R_{cd}. The design procedure utilizing the RSA coarse model R_c is the following:

1. Take initial design x^{init}.
2. Find the starting point $x^{(0)}$ for SM algorithm by optimizing the coarse-discretization model R_{cd}.

3. Allocate N base designs, $X_B = \{x^1, \ldots, x^N\}$.
4. Evaluate $\boldsymbol{R}_{cd}$ at each design x^j, $j = 1, 2, \ldots, N$.
5. Build the coarse model $\boldsymbol{R}_c$ as a response surface approximation of the data pairs $\{(x^j, \boldsymbol{R}_{cd}(x^j))\}_{j=1,\ldots,N}$.
6. Set $i = 0$.
7. Evaluate the fine model $\boldsymbol{R}_f$ at $x^{(i)}$.
8. Construct the surrogate model $\boldsymbol{R}_s^{(i)}$ as in (4.10) and (4.11).
9. Find a new design $x^{(i+1)}$ by optimizing $\boldsymbol{R}_s^{(i)}$ as in (3.1).
10. Set $i = i + 1$.
11. If the termination condition is not satisfied, go to 7.
12. End.

The first phase of the design process is to find an optimized design of the coarse-discretization model. The optimum of $\boldsymbol{R}_{cd}$ is usually the best design we can get at a reasonably low computational cost. This cost can be further reduced by relaxing tolerance requirements while searching for $x^{(0)}$: due to a limited accuracy of $\boldsymbol{R}_{cd}$, it is sufficient to find only a rough approximation of its optimum. Steps 3–5 describe the construction of the RSA coarse model. A popular choice of RSA is kriging interpolation (Queipo et al. 2005). Allocation of the base points is usually executed using Latin hypercube sampling (Beachkofski and Grandhi 2002). Steps 6–12 describe the flow of the SM algorithm. Figure 4.1 shows the flowchart of the design process.

4.3 Shape-Preserving Response Prediction

Shape-preserving response prediction (SPRP) (Koziel 2010a) is one of the recent SBO techniques that exploit the knowledge embedded in the low-fidelity model in order to predict the high-fidelity model response. Unlike space mapping, SPRP does not use any extractable parameters, and, therefore, it is more suitable for antenna design.

4.3.1 SPRP Concept

SPRP uses the generic surrogate-based optimization scheme (3.1). To construct the surrogate model, SPRP assumes that the change of the high-fidelity model response due to adjustments of the design variables can be predicted using the actual changes of the low-fidelity model response. It is important that the low-fidelity model is physics based so that the effect of the design parameter variations on the model response is similar for both models. In the context of antenna design, this property is generally ensured by using coarse-discretization low-fidelity models evaluated using the same EM solver as for the high-fidelity models.

In SPRP, the change of the low-fidelity model response is described by translation vectors corresponding to a certain (finite) number of characteristic points of the model's response. These translation vectors are subsequently used to predict the

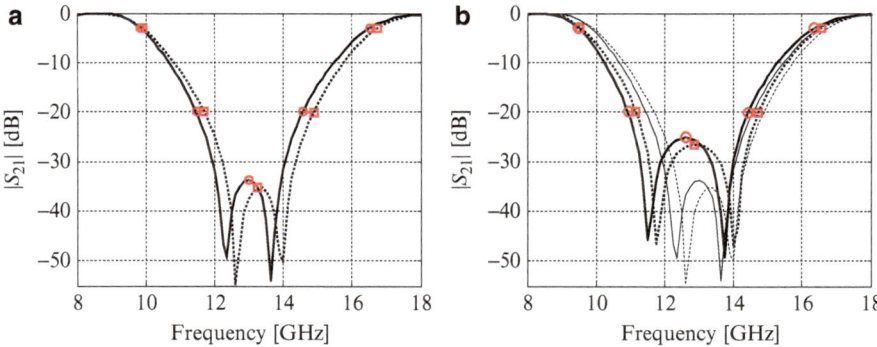

Fig. 4.3 SPRP concept: (**a**) Example low-fidelity model response at the design $x^{(i)}$, $R_c(x^{(i)})$ (*solid line*), the low-fidelity model response at x, $R_c(x)$ (*dotted line*), characteristic points of $R_c(x^{(i)})$ (*open circle*) and $R_c(x)$ (*open square*), and the translation vectors (*short lines*); (**b**) high-fidelity model response at $x^{(i)}$, $R_f(x^{(i)})$ (*solid line*) and the predicted high-fidelity model response at x (*dotted line*) obtained using SPRP based on characteristic points of (**a**); characteristic points of $R_f(x^{(i)})$ (*open circle*) and the translation vectors (*short lines*) were used to find the characteristic points (*open square*) of the predicted high-fidelity model response; low-fidelity model responses $R_c(x^{(i)})$ and $R_c(x)$ are plotted using thin solid and dotted line, respectively (Koziel 2010a)

change of the high-fidelity model response with the actual response of R_f at the current iteration point, $R_f(x^{(i)})$, treated as a reference.

Figure 4.3a shows an example low-fidelity model response, $|S_{21}|$ in the frequency range 8–18 GHz, at the design $x^{(i)}$, as well as the low-fidelity model response at some other design x. The responses are of a double folded stub bandstop filter example considered in Koziel (2010a). Circles denote the characteristic points of $R_c(x^{(i)})$, selected here to represent $|S_{21}| = -3$ dB, $|S_{21}| = -20$ dB, and the local $|S_{21}|$ maximum (at about 13 GHz). Squares denote the corresponding characteristic points for $R_c(x)$, while the line segments represent the translation vectors ("shift") of the characteristic points of R_c when changing the design variables from $x^{(i)}$ to x. Since the low-fidelity model is physics based, the high-fidelity model response at the given design, here, x, can be predicted using the same translation vectors applied to the corresponding characteristic points of the high-fidelity model response at $x^{(i)}$, $R_f(x^{(i)})$. This is illustrated in Fig. 4.3b.

4.3.2   SPRP Formulation

Rigorous formulation of SPRP uses the following notation concerning the responses: $R_f(x) = [R_f(x,\omega_1) \dots R_f(x,\omega_m)]^T$ and $R_c(x) = [R_c(x,\omega_1) \dots R_c(x,\omega_m)]^T$, where ω_j, $j = 1, \dots, m$, is the frequency sweep. Let $p_j^f = [\omega_j^f \; r_j^f]^T$, $p_j^{c0} = [\omega_j^{c0} \; r_j^{c0}]^T$, and $p_j^c = [\omega_j^c \; r_j^c]^T$, $j = 1, \dots, K$, denote the sets of characteristic points of $R_f(x^{(i)})$, $R_c(x^{(i)})$, and $R_c(x)$, respectively. Here, ω and r denote the frequency and magnitude components of the

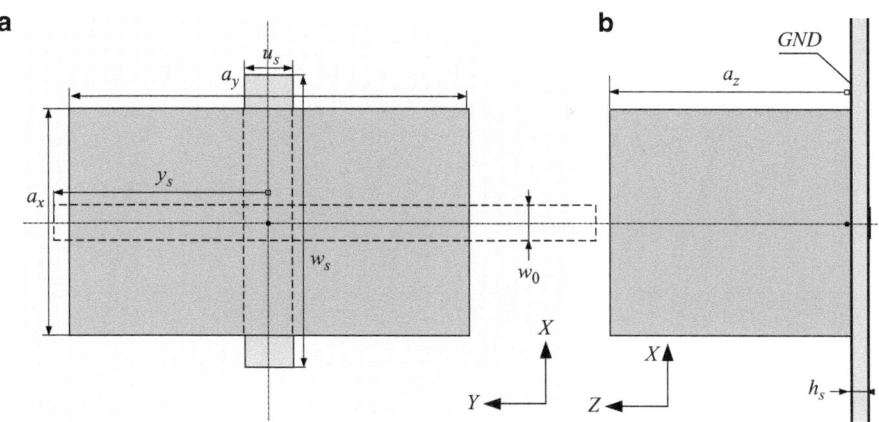

Fig. 4.4 DRA (Petosa 2007): (**a**) top and (**b**) side views

respective point. The translation vectors of the low-fidelity model response are defined as $t_j = [\omega_j^t \ r_j^t]^T$, $j = 1, \ldots, K$, where $\omega_j^t = \omega_j^c - \omega_j^{c0}$ and $r_j^t = r_j^c - r_j^{c0}$.

The SPRP surrogate model is defined as follows:

$$R_s^{(i)}(x) = \left[R_s^{(i)}(x, \omega_1) \quad \ldots \quad R_s^{(i)}(x, \omega_m) \right]^T \tag{4.12}$$

where

$$R_s^{(i)}(x, \omega_j) = R_{f,i}\left(x^{(i)}, F\left(\omega_j, \left\{-\omega_k^t\right\}_{k=1}^K\right)\right) + R\left(\omega_j, \left\{r_k^t\right\}_{k=1}^K\right) \tag{4.13}$$

for $j = 1, \ldots, m$. $R_{f,i}(x, \omega_1)$ is an interpolation of $\{R_f(x, \omega_1), \ldots, R_f(x, \omega_m)\}$ onto the frequency interval $[\omega_1, \omega_m]$.

The scaling function F interpolates the data pairs $\{\omega_1, \omega_1\}$, $\{\omega_1^f, \omega_1^f - \omega_1^t\}$, ..., $\{\omega_K^f, \omega_K^f - \omega_K^t\}$, $\{\omega_m, \omega_m\}$, onto the frequency interval $[\omega_1, \omega_m]$. The function R does a similar interpolation for data pairs $\{\omega_1, r_1\}$, $\{\omega_1^f, r_1^f - r_1^t\}$, ..., $\{\omega_K^f, r_K^f - r_K^t\}$, $\{\omega_m, r_m\}$; here $r_1 = R_c(x, \omega_1) - R_c(x^r, \omega_1)$ and $r_m = R_c(x, \omega_m) - R_c(x^r, \omega_m)$. In other words, the function F translates the frequency components of the characteristic points of $R_f(x^{(i)})$ to the frequencies at which they should be located according to the translation vectors t_j, while the function R adds the necessary magnitude component.

4.3.3 Illustration Example

The use of SPRP for antenna design will be demonstrated in further chapters. Here, we indicate the predictive power of the SPRP surrogate using an example of a dielectric resonator antenna (DRA) shown in Fig. 4.4. The DRA is installed at a ground plane and operates at the TEδ_{11} mode (Petosa 2007); see Fig. 4.4 for its

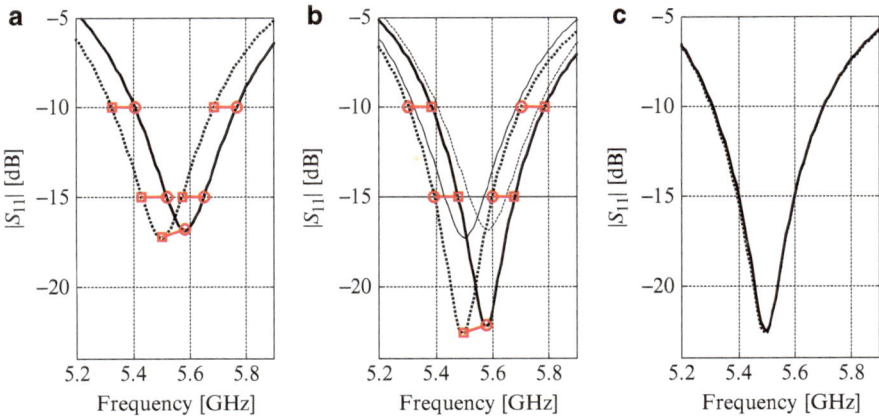

Fig. 4.5 SPRP surrogate for a DRA: (**a**) Low-fidelity model response at certain design $x^{(i)}$, $R_c(x^{(i)})$ (*solid line*), the low-fidelity model response at x, $R_c(x)$ (*dotted line*), characteristic points of $R_c(x^{(i)})$ (*open circle*) and $R_c(x)$ (*open square*), and the translation vectors (*short lines*); (**b**) high-fidelity model response at $x^{(i)}$, $R_f(x^{(i)})$ (*solid line*) and the predicted high-fidelity model response at some other design x (*dotted line*) obtained using SPRP based on characteristic points of Fig. 4.4a; characteristic points of $R_f(x^{(i)})$ (*open circle*) and the translation vectors (*short lines*) were used to find the characteristic points (*open square*) of the predicted high-fidelity model response; low-fidelity model responses $R_c(x^{(i)})$ and $R_c(x)$ are plotted using *thin solid* and *dotted line*, respectively; (**c**) actual and SPRP-predicted high-fidelity model response at x (*dotted line*)

geometry. The DRA is fed with a 50 Ω microstrip through a slot made in the ground plane. The design variables are $x=[a_x \, a_y \, a_z \, a_{y0} \, u_s \, w_s \, y_s]^T$. Relative permittivity and loss tangent of the DR core are 10 and 1e−4 respectively. Substrate is 0.5 mm thick RO4003C material (RO4000 2010). The width of the microstrip signal trace is 1.17 mm. Metallization of the trace and ground is with 50 μm copper.

The high-fidelity model R_f is simulated using the CST MWS transient solver (CST 2013) (~500,000 mesh cells, simulation time 11 min using a 2.66 GHz quad-core CPU with 4 GB RAM computer). The low-fidelity model R_c is also evaluated in CST but with coarser discretization (~15,000 mesh cells, evaluation time 24 s using the same computer). Figure 4.5 shows the responses of the high- and low-fidelity model of the DRA at a certain reference design, the construction of the SPRP surrogate, and the agreement between the SPRP-predicted and the actual high-fidelity model response.

4.3.4 Practical Issues

It should be emphasized that the physics-based low-fidelity model is critical for the method's performance. On the other hand, SPRP can be characterized as a nonparametric, nonlinear, and design-variable-dependent response correction. Its important feature is that SPRP does not use any extractable parameters which are normally found by solving a separate nonlinear minimization problem.

This SPRP property makes it suitable for antenna design because of a smaller number of low-fidelity model evaluations required in the optimization process as compared to, e.g., space mapping. It should be reiterated that the one-to-one correspondence between the characteristic points of the high- and low-fidelity model is fundamental for SPRP. If it is not satisfied, the SPRP surrogate will not be well defined and the entire method will not work. In some cases, this limitation can be alleviated by generalizations of SPRP, where the sets of corresponding characteristic points are generated based not on distinctive features of the responses (e.g., characteristic response levels or local minima/maxima) but by introducing additional points that are equally spaced in frequency and inserted between well-defined points (Koziel 2010a). These additional points not only ensure that the SPRP model (4.12), (4.13) is well defined but also allow us to capture the response shape of the models even though the number of distinctive features (e.g., local maxima and minima) is different for high- and low-fidelity models.

4.4   Adaptive Response Correction

Adaptive response correction (ARC) (Koziel et al. 2009b) is a generalization of the simple response correction technique (output space mapping; Bandler et al. 2004a, b). It has recently been applied to antenna design (Koziel and Ogurtsov 2013a). Similarly to SPRP, it does not use any extractable parameters. ARC does not rely on the assumptions required by SPRP regarding the similarity between the high- and low-fidelity model responses (i.e., correspondence between the sets of characteristic points, cf. Sect. 4.3).

The ARC surrogate $R^{(i)}_{s \cdot ARC}$ is defined as

$$R^{(i)}_{s \cdot ARC}\left(x \right) = R_c \left(x \right) + d_{ARC}\left(x, x^{(i)} \right) \tag{4.14}$$

where $d_{ARC}(x, x^{(i)})$ is the response correction term dependent on the design variables x. We want to maintain a perfect match between R_f and the surrogate at $x^{(i)}$, i.e., $d_{ARC}(x, x^{(i)})$ must satisfy

$$d_{ARC}\left(x^{(i)}, x^{(i)} \right) = d^{(i)} = R_f \left(x^{(i)} \right) - R_c \left(x^{(i)} \right) \tag{4.15}$$

with $d^{(i)}$ being a basic response correction term (output SM; Bandler et al. 2004a, b). The idea behind ARC is to account for the difference between $R_c(x)$ and $R_c(x^{(i)})$ and to modify the correction term $d^{(i)}$ so that this modification reflects changes of R_c during the surrogate optimization: if R_c shifts or changes its shape with respect to frequency, ARC should track these changes.

Figure 4.6 shows R_f and R_c at two different designs and the corresponding additive correction terms d (4.15). The relation between these terms is similar to the relation between the low-fidelity model responses so that tracking changes of R_f help to determine necessary changes to $d^{(i)}$. For the purpose of the ARC formulation, we can consider the explicit dependence of the model responses on frequency ω, so that

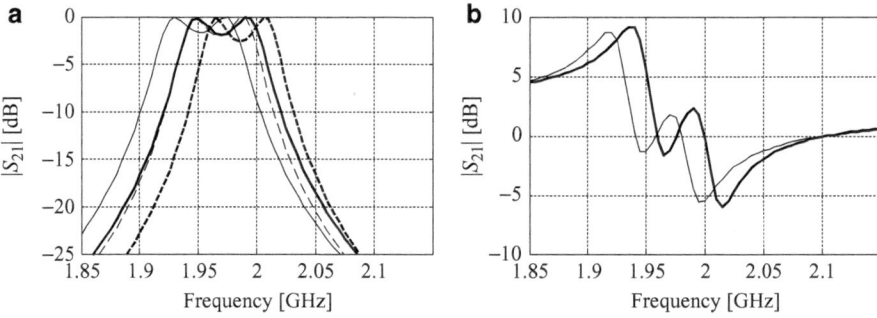

Fig. 4.6 (**a**) R_f (*solid line*) and R_c (*dashed line*) at certain design, and R_f (*thick solid line*) and R_c (*dashed line*) at a different design; (**b**) additive correction terms corresponding to responses of (**a**)

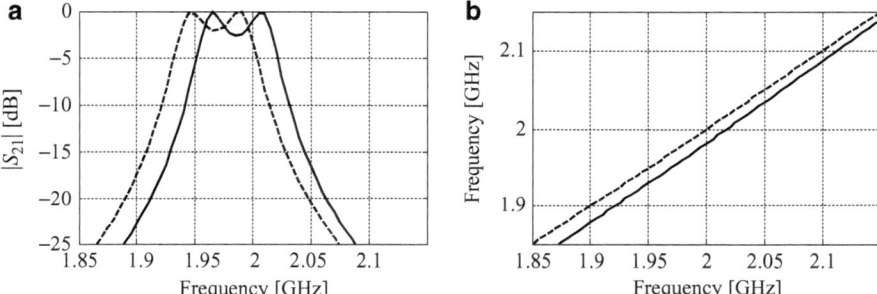

Fig. 4.7 (**a**) Low-fidelity model responses $R_c(x^{(i)})$ (*dashed line*) and $R_c(x)$ (*solid line*); (**b**) mapping function $F^{(i)}(x)$ (*solid line*) obtained as described in the text, and the identity function (which equals to $F^{(i)}(x^{(i)})$) (*dashed line*)

$R_c(x,\omega)$ is the value of $R_c(x)$ at ω. The core of ARC is a function $F^{(i)}$: $X \times \Omega \to [\omega_{min}, \omega_{max}]$, where X stands for the low- and high-fidelity model domain, Ω is a frequency band of interest, and $[\omega_{min}, \omega_{max}] \supseteq \Omega$ is its possible expansion. $F^{(i)}$ is established at iteration i so that the difference between $R_c(x,\omega)$ and $R_c(x^{(i)}, F^{(i)}(x,\omega))$ (the F-scaled $R_c(x,\omega)$) is minimized in the L-square sense, i.e., $\|R_c(x^{(i)}, F^{(i)}(x,\omega)) - R_c(x,\omega)\|$. Thus, $F^{(i)}$ is supposed to be defined so that mapped frequency $F^{(i)}(x,\omega)$ reflects the change of the R_c response at x with respect to $x^{(i)}$.

The ARC correction term is defined as follows (here, the dependence of $d_{ARC}(x, x^{(i)})$ on ω is shown explicitly):

$$d_{ARC}\left(x, x^{(i)}, \omega\right) = R_f\left(x^{(i)}, F^{(i)}\left(x, \omega\right)\right) - R_c\left(x^{(i)}, F^{(i)}\left(x, \omega\right)\right) \qquad (4.16)$$

Here, $F^{(i)}$ is implemented using a third-order polynomial; however, other realizations are also possible (Koziel et al. 2009b). Figures 4.7 and 4.8 illustrate the operation of the ARC technique. It can be observed that ARC gives much more accurate prediction of the high-fidelity model response than the simple response correction (4.15).

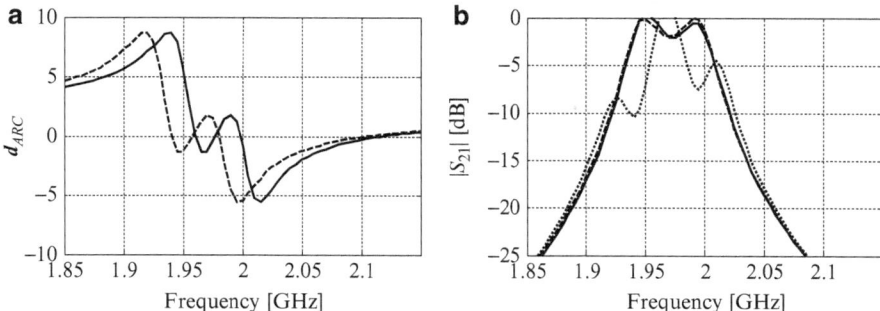

Fig. 4.8 (**a**) ARC correction terms $d_{ARC}(x^{(i)}, x^{(i)})$ (*dashed line*) and $d_{ARC}(x, x^{(i)})$ (*solid line*); (**b**) predicted R_f response at $x^{(i)}$ obtained with ARC, i.e., $R_{s-ARC}^{(i)}(x) = R_c(x) + d_{ARC}(x, x^{(i)})$ (*solid line*), actual R_f response $R_f(x)$ (*dashed line*), and predicted high-fidelity model response $R_s^{(i)}(x) = R_c(x) + d^{(i)}$ (*dotted line*)

In comparison with the conventional, additive response correction defined using the constant vector $d^{(i)} = R_f(x^{(i)}) - R_c(x^{(i)})$, the adaptive response correction is capable to better utilize the knowledge about the antenna structure contained in the low-fidelity model R_c, which leads to the smaller number of iterations (and, thus, lower computational cost) necessary to find the optimized design. This was demonstrated in Koziel et al. (2009a, b) for microwave filters and in Koziel and Ogurtsov (2013a) for antenna structures.

4.5 Manifold Mapping

Manifold mapping (MM) (Echeverria and Hemker 2005; Echeverria 2007a, b) is a simple, yet efficient SBO technique, which can be considered as a specialized case of output space mapping. MM is supported by the rigorous convergence theory (Echeverría and Hemker 2008; Echeverria 2007a). Like many other SBO methods, manifold mapping uses the generic scheme (3.1). The MM surrogate model $R_s^{(i)}$ is defined as

$$R_s^{(i)}(x) = R_f\left(x^{(i)}\right) + S^{(i)}\left(R_c(x) - R_c\left(x^{(i)}\right)\right), \tag{4.17}$$

where $S^{(i)}$, for $i \geq 1$, is the $m \times m$ matrix given by

$$s^{(i)} = \Delta F \Delta C^\dagger, \tag{4.18}$$

with

$$\Delta F = \left[R_f\left(x^{(i)}\right) - R_f(x^{(i-1)}) \quad \cdots \quad R_f\left(x^{(i)}\right) - R_f(x^{(\max\{i-n,0\})}) \right], \tag{4.19}$$

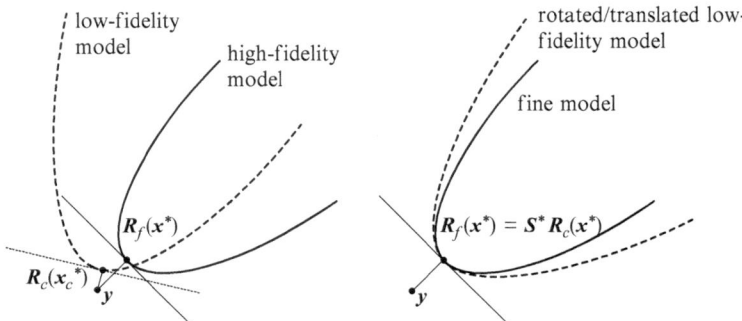

Fig. 4.9 Illustration of the manifold-mapping model alignment for a least-squares optimization problem. The point x_c^* denotes the minimizer corresponding to the coarse model response, and the point y is the vector of design specifications. *Thin solid* and *dashed straight lines* denote the tangent planes for the fine and coarse model response at their optimal designs, respectively. By the linear correction S^*, the point $R_c(x^*)$ is mapped to $R_f(x^*)$ and the tangent plane for $R_c(x)$ at $R_c(x^*)$ to the tangent plane for $R_f(x)$ at $R_f(x^*)$ (Koziel et al. 2011c)

$$\Delta C = \left[R_c\left(x^{(i)}\right) - R_c\left(x^{(i-1)}\right) \quad \cdots \quad R_c\left(x^{(i)}\right) - R_c\left(x^{(\max\{i-n,0\})}\right) \right]. \quad (4.20)$$

The matrix $S^{(0)}$ is typically taken as the identity matrix I_m. Here, † denotes the pseudoinverse operator defined for ΔC as

$$\Delta C^{\dagger} = V_{\Delta C} \Sigma_{\Delta C}^{\dagger} U_{\Delta C}^{T}, \quad (4.21)$$

where $U\Delta C$, $\Sigma\Delta C$, and $V\Delta C$ are the factors in the singular value decomposition of ΔC. The matrix $\Sigma \Delta C^{\dagger}$ is the result of inverting the nonzero entries in $\Sigma \Delta C$, leaving the zeroes invariant (Echeverria and Hemker 2005). Some mild general assumptions on the model responses are made in theory (Echeverria 2007a) so that every pseudoinverse introduced is well defined.

The response correction $R_s^{(i)}(x)$ is an approximation of

$$R_s^*(x) = R_f\left(x^*\right) + S^*\left(R_c(x) - R_c(x^*)\right), \quad (4.22)$$

with S^* being the $m \times m$ matrix defined as

$$S^* = J_f\left(x^*\right) J_c\left(x^*\right), \quad (4.23)$$

where $J_f(x^*)$ and $J_c(x^*)$ stand for the fine and coarse model response Jacobian, respectively, evaluated at x^*. Obviously, neither x^* nor S^* is known beforehand. Therefore, one needs to use an iterative approximation, such as the one in (4.17)–(4.21), in the actual manifold-mapping algorithm.

Illustration of the manifold-mapping model alignment is presented in Fig. 4.9 for the least-squares optimization problem $U(R_f(x)) = \|R_f(x) - y\|_2^2$ with $y \in R^m$ being the design specifications given. In that figure the point x_c^* denotes the minimizer corresponding to the coarse model cost function $U(\boldsymbol{R}_c(\boldsymbol{x}))$. In case of unconstrained problem, the optimality associated to the least-squares objective function is translated into the orthogonality between the tangent plane for $\boldsymbol{R}_f(\boldsymbol{x})$ at $\boldsymbol{x}^*$ and the vector $\boldsymbol{R}_f(\boldsymbol{x}^*) - \boldsymbol{y}$.

For least-squares optimization problems, manifold mapping is supported by mathematically sound convergence theory (Echeverria and Hemker 2005). In general, manifold-mapping algorithms can be expected to converge for a merit function U sufficiently smooth. Since the correction in (4.17) does not involve U, if the model responses are smooth enough, and even when U is not differentiable, manifold mapping may still yield satisfactory solutions. The experimental evidence given in Koziel and Echeverria (2010) for designs based on minimax objective functions indicates that the MM approach can be used successfully in more general situations than those for which theoretical results have been obtained. Various modifications and enhancements of the basic manifold-mapping algorithm presented here have been proposed in the literature (Echeverría and Hemker 2008; Hemker and Echeverría 2007), including incorporation of the convergence safeguards analogous to a trust-region method (Conn et al. 2000).

4.6 Adaptively Adjusted Design Specifications

In order to realize efficient surrogate-based optimization process, it is not necessary to remove the discrepancies between the low- and high-fidelity models by correcting the latter. Another way is to "absorb" the model misalignment by proper adjustment of the design specifications. In microwave engineering in general and in antenna design in particular, most of the design tasks can be formulated as minimax problems with upper and lower specifications, and it is easy to implement modifications by, for example, shifting the specification levels and corresponding frequency bands. This approach, both easy to implement and efficient, is exploited by adaptively adjusted design specifications (AADS) technique (Koziel 2010b) described in this section. AADS consists of the following two simple steps that can be iterated if necessary:

1. Modify the original design specifications in order to take into account the difference between the responses of $\boldsymbol{R}_f$ and $\boldsymbol{R}_c$ at their characteristic points.
2. Obtain a new design by optimizing the low-fidelity model with respect to the modified specifications.

Characteristic points of the responses should correspond to the design specification levels. They should also include local maxima/minima of the respective responses at which the specifications may not be satisfied. Figure 4.10b shows characteristic points of $\boldsymbol{R}_f$ and $\boldsymbol{R}_c$ for our bandstop filter example. The points correspond to −3 and −30 dB levels as well as to the local maxima of the responses. As one can observe in Fig. 4.10b, the selection of points is rather straightforward.

In the first step of the optimization procedure, the design specifications are modified so that the level of satisfying/violating the modified specifications by the

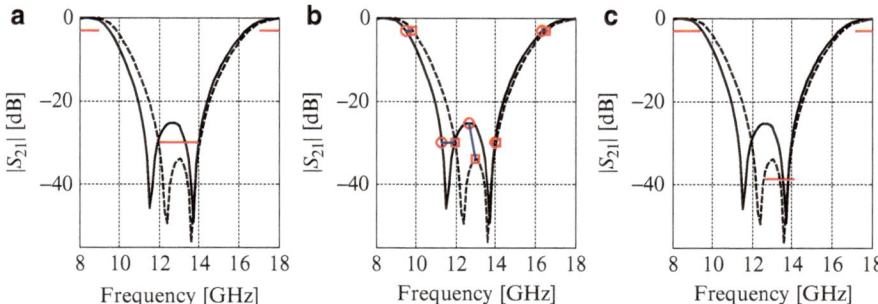

Fig. 4.10 Bandstop filter example (responses of R_f and R_c are denoted using *solid* and *dashed line*, respectively) (Koziel 2010b): (**a**) responses at the initial design (low-fidelity model optimum) as well as the original design specifications, (**b**) characteristic points of the responses corresponding to the specification levels (here, −3 and −30 dB) and to the local maxima, (**c**) responses at the initial design as well as the modified design specifications

low-fidelity model response corresponds to the satisfaction/violation levels of the original specifications by the high-fidelity model response. More specifically, for each edge of the specification line, the edge frequency is shifted by the difference of the frequencies of the corresponding characteristic points, e.g., the left edge of the specification line of −30 dB is moved to the right by about 0.7 GHz, which is equal to the length of the line connecting the corresponding characteristic points in Fig. 4.10b. Similarly, the specification levels are shifted by the difference between the local maxima/minima values for the respective points, e.g., the −30 dB level is shifted down by about 8.5 dB because of the difference of the local maxima of the corresponding characteristic points of R_f and R_c. Modified design specifications are shown in Fig. 4.10c.

The low-fidelity model is subsequently optimized with respect to the modified specifications and the new design obtained this way is treated as an approximated solution to the original design problem (i.e., optimization of the high-fidelity model with respect to the original specifications). Steps 1 and 2 can be repeated if necessary. As indicated in Koziel (2010b), substantial design improvement is typically observed after the first iteration; however, additional iterations may bring further enhancement. In practice, the algorithm is terminated once the current iteration does not bring further improvement of the high-fidelity model design.

It should be emphasized that unlike in the case of other simulation-driven techniques popular in microwave engineering (particularly space mapping; Bandler et al. 2004a, b), in AADS, the low-fidelity model is not modified or corrected in any way. The discrepancy between the models is "absorbed" by means of modifying the design specifications.

The operation of the adaptively adjusted design specifications technique can probably be best explained using the example. Figure 4.11 illustrates an iteration of the procedure used for design of a CBCPW-to-SIW transition (Koziel 2011). One can observe that the absolute matching between the low- and high-fidelity

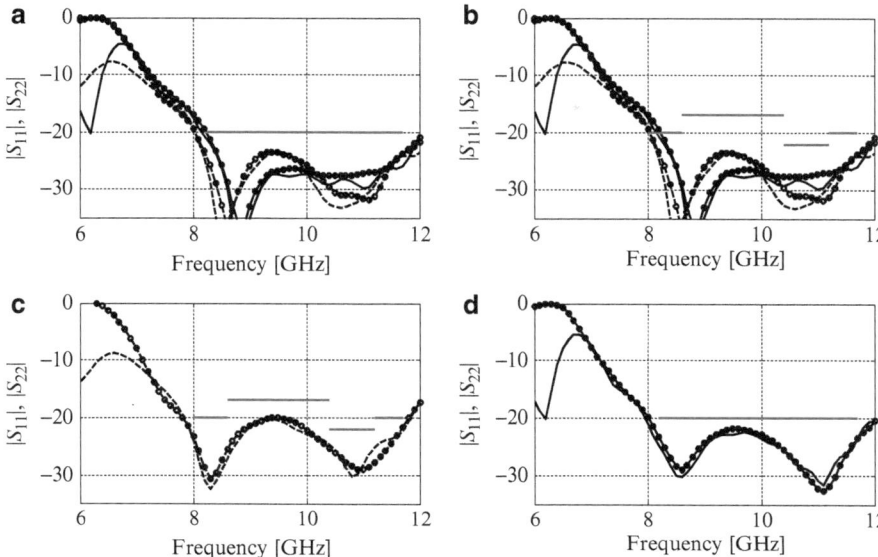

Fig. 4.11 Adaptively adjusted design specification technique applied to optimize CBCPW-to-SIW transitions (Koziel 2011). R_f and R_c responses are denoted as *solid* and *dashed lines*, respectively. $|S_{22}|$ distinguished from $|S_{11}|$ using *circles*. Design specifications denoted by *thick horizontal lines*. (**a**) R_f and R_c responses at the beginning of the iteration as well as original design specifications; (**b**) R_f and R_c responses and modified design specifications that reflect the differences between the responses; (**c**) low-fidelity model optimized to meet the modified specifications; (**d**) high-fidelity model at the low-fidelity model optimum shown versus original specifications. *Horizontal lines* indicate the design specifications

models is not as important as the shape similarity. It should be stressed that the low-fidelity model is not modified in any way, that is, no changes are applied to it in order to align it with the high-fidelity model. The discrepancy between the high- and low-fidelity model responses is accounted for by modifying the design specifications.

4.7   Multi-fidelity Design Optimization

One of the most robust physics-based SBO techniques exploiting coarse-discretization EM simulations is a multi-fidelity design optimization algorithm (Koziel and Ogurtsov 2010b). It is simple to implement, and, except the last state (design refinement), it does not require any modification of the low-fidelity models.

The multi-fidelity design optimization methodology is based on a family of coarse-discretization models $\{R_{c\cdot j}\}, j = 1, \ldots, K$, all evaluated by the same EM solver as the one used for the high-fidelity model. Discretization of the model $R_{c\cdot j+1}$ is finer than that of the model $R_{c\cdot j}$, which results in better accuracy but also longer evaluation time. In practice, the number of coarse-discretization models is two or three.

In multi-fidelity algorithm, the design procedure starts from optimizing the lowest-fidelity model $R_{c\cdot 1}$ starting from the initial design $x^{(0)}$. The design $x^{(1)}$ obtained this way becomes a starting point for optimizing the next model, $R_{c\cdot 2}$. The process continues until all coarse-discretization models are optimized. Having the approximate optimum $x^{(K)}$ of the last (and finest) coarse-discretization model $R_{c\cdot K}$, the model $R_{c\cdot K}$ is evaluated at all perturbed designs around $x^{(K)}$, i.e., at $x_k^{(K)} = [x_1^{(K)} \ldots x_k^{(K)} + \mathrm{sign}(k)\cdot d_k \ldots x_n^{(K)}]^T$, $k = -n, -n+1, \ldots, n-1, n$. We use the following notation: $R^{(k)} = R_{c\cdot K}(x_k^{(K)})$. This data can be used to refine the final design without directly optimizing R_f. Instead, an approximation model involving $R^{(k)}$ is set up and optimized in the neighborhood of $x^{(K)}$ defined as $[x^{(K)} - d, x^{(K)} + d]$, where $d = [d_1 \, d_2 \ldots d_n]^T$. The size of the neighborhood can be selected based on sensitivity analysis of $R_{c\cdot 1}$ (the cheapest of the coarse-discretization models); usually d equals a few percent of $x^{(K)}$.

Here, the approximation is performed using a reduced quadratic model $q(x) = [q_1 \, q_2 \ldots q_m]^T$, defined as

$$q_j(x) = q_j\left([x_1 \ldots x_n]^T\right) = \lambda_{j\cdot 0} + \lambda_{j\cdot 1}x_1 + \cdots + \lambda_{j\cdot n}x_n + \lambda_{j\cdot n+1}x_1^2 + \cdots + \lambda_{j\cdot 2n}x_n^2 \quad (4.24)$$

Coefficients $\lambda_{j\cdot r}$, $j = 1, \ldots, m$, $r = 0, 1, \ldots, 2n$, can be uniquely obtained by solving the linear regression problems

$$\begin{bmatrix} 1 & x_{-n\cdot 1}^{(K)} & \cdots & x_{-n.n}^{(K)} & \left(x_{-n\cdot 1}^{(K)}\right)^2 & \cdots & \left(x_{-n\cdot n}^{(K)}\right)^2 \\ \vdots & \vdots & & \vdots & & \vdots & \vdots \\ 1 & x_{0\cdot 1}^{(K)} & \cdots & x_{0\cdot n}^{(K)} & \left(x_{0\cdot 1}^{(K)}\right)^2 & \cdots & \left(x_{-n\cdot n}^{(K)}\right)^2 \\ \vdots & \vdots & & \vdots & & \vdots & \vdots \\ 1 & x_{n\cdot 1}^{(K)} & \cdots & x_{n\cdot n}^{(K)} & \left(x_{n\cdot 1}^{(K)}\right)^2 & \cdots & \left(x_{-n\cdot n}^{(K)}\right)^2 \end{bmatrix} \cdot \begin{bmatrix} \lambda_{j\cdot 0} \\ \lambda_{j\cdot 1} \\ \vdots \\ \lambda_{j\cdot 2n} \end{bmatrix} = \begin{bmatrix} R_j^{(-n)} \\ \vdots \\ R_j^{(0)} \\ \vdots \\ R_j^{(n)} \end{bmatrix} \quad (4.25)$$

where $x_{k\cdot j}^{(K)}$ is a jth component of the vector $x_k^{(K)}$ and $R_j^{(k)}$ is a jth component of the vector $R^{(k)}$.

In order to account for unavoidable misalignment between $R_{c\cdot K}$ and R_f, instead of optimizing the quadratic model q, it is recommended to optimize a corrected model $q(x) + [R_f(x^{(K)}) - R_{c\cdot K}(x^{(K)})]$ that ensures a zero-order consistency (Alexandrov and Lewis 2001) between $R_{c\cdot K}$ and R_f. The refined design can be then found as

$$x^* = \arg \min_{x^{(K)} - d \le x \le x^{(K)} + d} U\left(q(x) + \left[R_f\left(x^{(K)}\right) - R_{c\cdot K}\left(x^{(K)}\right)\right]\right) \quad (4.26)$$

This kind of correction is also known as output space mapping (Koziel et al. 2006). If necessary, the step (4.26) can be performed a few times starting from a refined design, i.e., $x^* = \mathrm{argmin}\{x^{(K)} - d \le x \le x^{(K)} + d: U(q(x) + [R_f(x^*) - R_{c\cdot K}(x^*)])\}$ (each iteration requires only one evaluation of R_f).

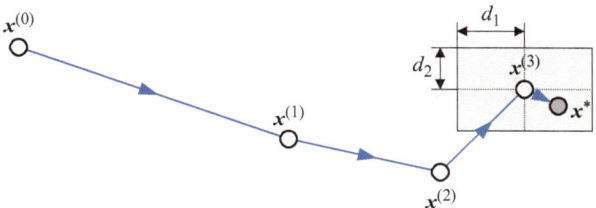

Fig. 4.12 Operation of the multi-fidelity design optimization procedure for $K=3$ (three coarse-discretization models). The design $x^{(j)}$ is obtained as the optimal solution of the model $R_{c \cdot j}, j = 1, 2, 3$. A reduced second-order approximation model q is set up in the neighborhood of $x^{(3)}$ (*gray area*) and the final design x^* is obtained by optimizing a response surface approximation model q as in (4.26)

The design optimization procedure can be summarized as follows (input arguments are initial design $x^{(0)}$ and the number of coarse-discretization models K):

1. Set $j=1$.
2. Optimize coarse-discretization model $R_{c \cdot j}$ to obtain a new design $x^{(j)}$ using $x^{(j-1)}$ as a starting point.
3. Set $j=j+1$; if $j<K$, go to 2.
4. Obtain a refined design x^* as in (4.26).
5. End.

Note that the original model R_f is only evaluated at the final stage (step 4) of the optimization process. Operation of the algorithm is illustrated in Fig. 4.12. Coarse-discretization models can be optimized using any available algorithm.

As mentioned above, the number K of coarse-discretization models is typically two or three. The first coarse-discretization model $R_{c \cdot 1}$ should be set up so that its evaluation time is at least 30–100 times shorter than the evaluation time of the fine model. The reason is that the initial design may be quite poor so that the expected number of evaluations of $R_{c \cdot 1}$ is usually large. By keeping $R_{c \cdot 1}$ fast, one can control the computational overhead related to its optimization. Accuracy of $R_{c \cdot 1}$ is not critical because its optimal design is only supposed to give a rough estimate of the fine model optimum. The second (and, possibly third) coarse-discretization model should be more accurate but still at least about ten times faster than the fine model. This can be achieved by proper manipulation of the solver mesh density.

Various modifications and generalizations of the multi-fidelity algorithm have been proposed in the literature that aim at improving the robustness and reducing the computational cost of the design process; see, e.g., Koziel and Ogurtsov (2013b, c).

Chapter 5
Low-Fidelity Antenna Models

In this chapter, we discuss low-fidelity models as fundamental components of surrogate-based antenna optimization. In particular, we consider general requirements imposed on such models as well as overview typical low-fidelity model set-ups. Numerical examples are also provided to illustrate the trade-off between low-fidelity model accuracy and computational cost.

5.1 Low-Fidelity Models in Simulation-Driven Optimization

Antenna design methods discussed in this book exploit physics-based surrogate models. As explained in Chap. 3, this type of surrogates utilizes underlying low-fidelity (or coarse) electromagnetic (EM) models which are typically simulated with discrete full-wave EM solvers, custom or commercial (e.g., HFSS 2010, CST Microwave Studio 2013, and FEKO 2011). Examples of antenna structures for which full-wave discrete simulation is the only modeling possibility include but are not limited to ultra-wideband (UWB) antennas (Schantz 2005), dielectric resonator antennas (DRAs) (Petosa 2007), and antenna arrays with strong element coupling (Balanis 2005).

In the design optimization process, the low-fidelity model is to be simulated multiple times, either at a separate stage to create an auxiliary response surface surrogate (Koziel and Ogurtsov 2011a) or as a part of the SBO algorithm run to yield a prediction of the high-fidelity model optimum (Bandler et al. 2004a, b). As a result, the overall computational cost of the low-fidelity model simulations can substantially contribute to the total optimization cost. Therefore, one of the issues of SBO-based antenna design is to ensure that the low-fidelity models are as fast as possible. On the other hand, we want the models be reliable so that they adequately represent—both qualitatively and quantitatively—the high-fidelity model of the antenna structure under consideration over the simulation bandwidth. While establishing the low-fidelity model, one has to trade its accuracy for its computational speed.

S. Koziel and S. Ogurtsov, *Antenna Design by Simulation-Driven Optimization*,
SpringerBriefs in Optimization, DOI 10.1007/978-3-319-04367-8_5,
© Slawomir Koziel and Stanislav Ogurtsov 2014

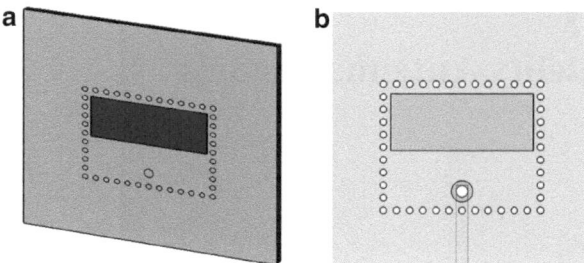

Fig. 5.1 Substrate integrated half-mode 5 GHz antenna at a particular design: (**a**) perspective view with the ground and substrate shown with finite lateral extends; (**b**) the *top view* with the top ground transparent

Subsequently, the tolerated inaccuracy of the low-fidelity model is corrected at the stage of the surrogate model update.

Another aspect of setting low-fidelity models for surrogate optimization of antennas is that we are often interested in both radiation and reflection responses. The responses, however, have different sensitivity to the coarseness of the model: the far-field quantities such as gain, radiation pattern, power radiated within the main beam, etc., are integral figures; therefore, they are less sensitive. On the other hand, the input impedance or the antenna reflection coefficient is more sensitive to fidelity of the antenna model, in particular, to how we describe the antenna feeding mechanism. Thus, our primary figures of interest in the process of optimization of a particular antenna affect possible simplifications introduced to create the low-fidelity model. In some cases, one can even disregard certain specific antenna responses in considerations or consider them in a separate stage of the design process; see, e.g., Koziel and Ogurtsov (2010a).

An example of a substrate integrated half-mode 5 GHz antenna shown in Fig. 5.1 illustrates differences in its responses evaluated with models of different fidelity as well as sensitivity of the antenna responses on the model fidelity. Both of the models are defined, discretized, and simulated using CST MWS (CST Microwave Studio 2013). The fine-model is discretized with about 451,000 mesh cells and evaluated in 69 min 26 s using a 2.33 GHz 8 core CPU with 8 GB RAM computer. A quite dense discretization of the model, which turns in a substantial simulation time, is a result of ensuring no feasible changes of the response versus discretization density. At the same time, the low-fidelity model contains only about 26,000 mesh cells, and its simulation time is only 26 s with the same computer. For this antenna—supposed to operate around 5 GHz—the reflection response of the low-fidelity model, shown in Fig 5.2a, is substantially misaligned with that of the high-fidelity one. The response of the latter indicates that antenna geometry needs to be tuned for 5 GHz operation. Therefore, the use of the low-fidelity model in the process of antenna adjustment can result in high reflection of a manufactured sample around 5 GHz, whereas the use of the low-fidelity model will be associated with a substantial total design time. Other figures of the antennas are not so sensitive to the model fidelity, as illustrated in Fig. 5.2c and d where we see no essential differences of the gain patterns of the two models.

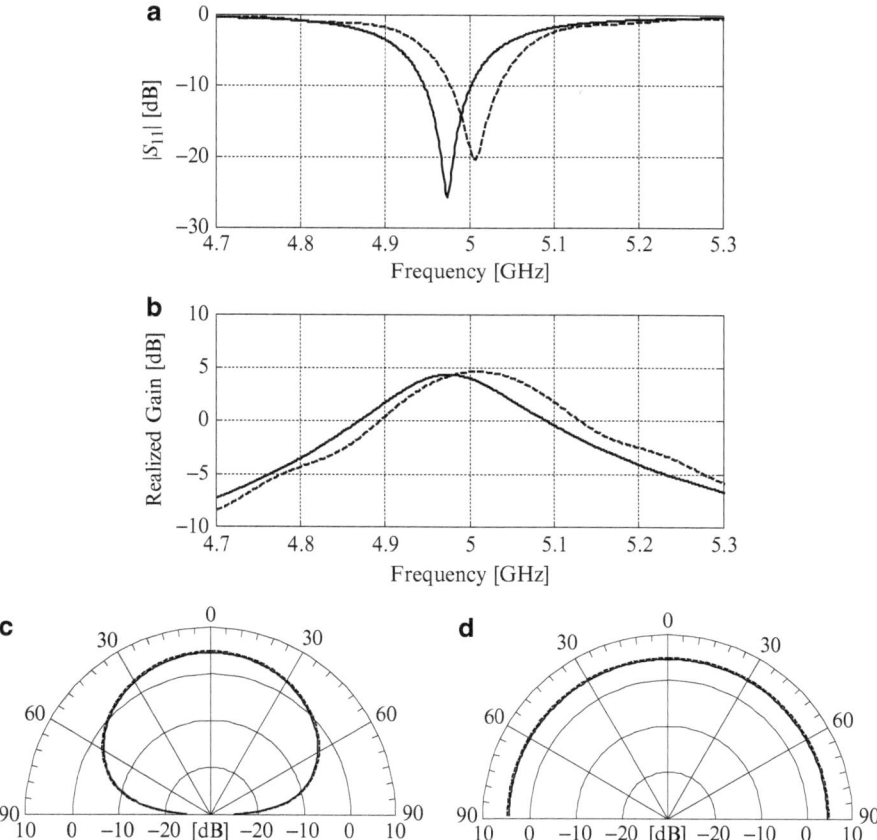

Fig. 5.2 Responses of the high-fidelity model (*solid line*) and low-fidelity model (*dash line*): (**a**) reflection, (**b**) realized gain for the zero zenith angle versus frequency, (**c**) IEEE gain at 5 GHz in the H-plane, and (**d**) IEEE gain at 5 GHz in the E-plane

5.2 Coarse-Discretization Antenna Models as a Basis for Low-Fidelity Antenna Models

Discrete EM simulators, in particular commercial software packages, e.g., HFSS (2010), CST Microwave Studio (2013), and FEKO (2011), are extensively used in the modern antenna design in both industry and academia (Kempel 2007). Not long time ago, discrete EM simulators were used mostly for design verification purposes. Nowadays, due to the progress in computing hardware as well as development of computational electromagnetic methods, the discrete EM simulators turn to be indispensable for the entire design process starting from a concept estimation step. The use of discrete full-wave simulators is also appealing from the practical point of view and allows obtaining reliable antenna responses with respect to environment and feeds.

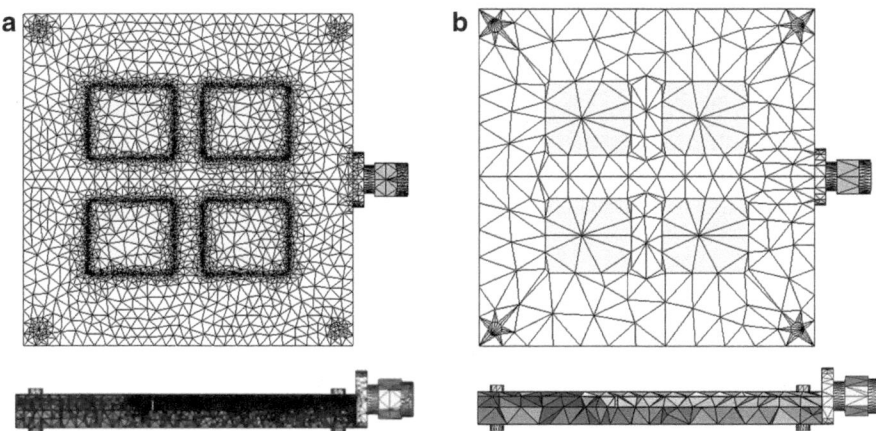

Fig. 5.3 Microstrip antenna (Chen 2008): (**a**) high-fidelity model shown with a fine tetrahedral mesh and (**b**) low-fidelity model shown with a much coarser mesh

With the discrete solvers, it is the discretization density that has the strongest impact on the accuracy and computational time of a particular antenna model. At the same time, the discretization density or mesh quality is probably the most efficient way to trade accuracy for speed. Therefore, a straightforward way to create a low-fidelity model of the antenna is through coarser mesh settings compared to those of the high-fidelity antenna model, e.g., as illustrated in Fig. 5.3. Because of possible simplifications, the low-fidelity model R_c is faster than R_f, typically it can be made 10–50 times faster; however, model R_c is obviously not as accurate as R_f. Therefore, the low-fidelity model cannot simply replace the high-fidelity model in the design optimization process. Figure 5.4 shows the high- and low-fidelity model responses at a specific design for the antenna of Fig. 5.3 obtained with different meshes, as well as the relationship between mesh coarseness and simulation time.

Selection of the model coarseness strongly affects the simulation time and performance of the design optimization process. Coarser models are faster, and it turns into a lower cost per design iteration while using SBO process (cf. (3.1)). The coarser models, however, are less accurate, which may result in a larger number of iterations necessary to find a satisfactory design. Also, there is an increased risk of failure for the optimization algorithm to find a good design (Koziel and Ogurtsov 2012b; see also Chap. 12 for more extensive discussion of this subject). Finer models, on the other hand, are more expensive but they are more likely to produce a useful design with a smaller number of iteration. One can infer from Fig. 5.4 that the two "finest" coarse-discretization models (with ~400,000 and ~740,000 mesh cells) represent the high-fidelity model response (shown as a thick solid line) quite properly. The model with ~270,000 cells can be considered as a borderline one. The two remaining models could be considered as poor ones, particularly the model with ~20,000 cells; its response is essentially unreliable.

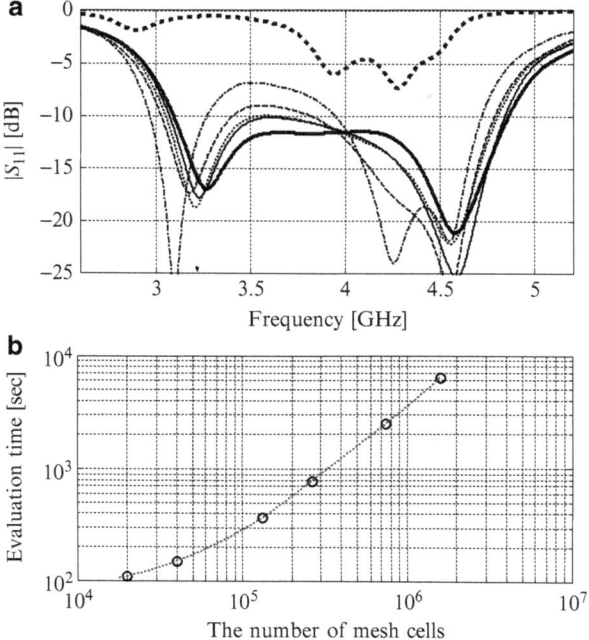

Fig. 5.4 Antenna of Fig. 5.3 at a selected design simulated with the CST MWS transient solver (CST Microwave Studio 2013): (**a**) reflection response of different discretization densities, 19,866 cells (*filled squares*), 40,068 cells (*dotted dashed line*), 266,396 cells (*dashed line*), 413,946 cells (*dotted line*), 740,740 cells (*solid line*), and 1,588,608 cells (*thick solid line*), and (**b**) the antenna simulation time versus the number of mesh cells (Koziel and Ogurtsov 2012b)

5.3   Additional Simplifications of Low-Fidelity Antenna Models

In addition to a coarser mesh, other simplifications can be made in the low-fidelity models. Possible computational simplifications include:

(a) Shrinking the computational domain and applying simple absorbing boundaries with the finite-volume methods implemented in the EM software in use (HFSS 2010, CST 2013, Taflove and Hagness 2006, and Lin 2002).

(b) Using low-order basis functions with the finite element and moment method solvers (Lin 2002, Harrington 1993, and Makarov 2002).

(c) Using more relaxed solution termination criteria such as the S-parameter error for the frequency domain methods with adaptive meshing (e.g., HFSS 2010 and CST Microwave Studio 2013) and residue energy for the time-domain solvers (CST Microwave Studio 2013).

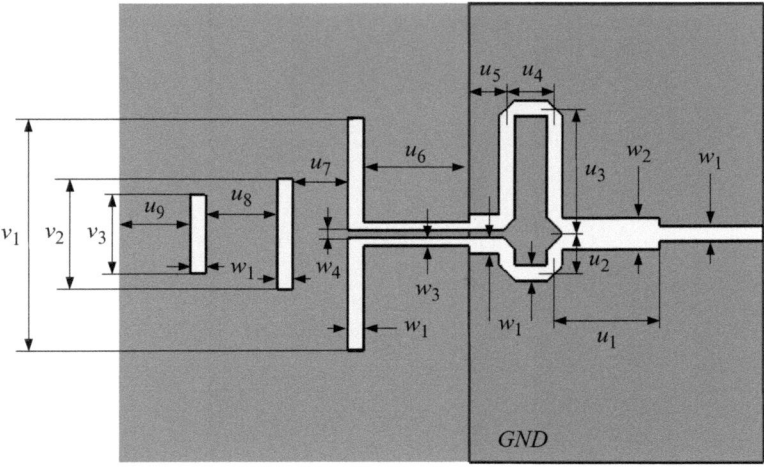

Fig. 5.5 Planar Yagi antenna. Substrate is shown semitransparent

Simplification of physics of the models can be the following:

(a) Ignoring dielectric and metal losses as well as material dispersion if their impact to the simulated response is not significant.
(b) Setting metallization thickness to zero for traces, strips, and patches.
(c) Ignoring moderate anisotropy of substrates.
(d) Energizing the antenna with discrete sources rather than waveguide ports (HFSS 2010, CST Microwave Studio 2013, and FEKO 2011 and Taflove and Hagness 2006).

Rigorously speaking, computational and physical simplifications are closely related and listed in two groups mostly for classification purposes. For example, ignoring dielectric losses and material dispersion in the model simulated with a time-domain finite-volume method turns into a much simpler formulation of the solution process with a smaller number of unknowns for the same mesh discretization (Taflove and Hagness 2006).

The following example illustrates the effect of the simplifications listed above on accuracy of the low-fidelity model response as well as on the model evaluation time. Consider a planar Yagi antenna shown in Fig. 5.5 (Deal et al 2000 and Kaneda et al 2002). The substrate is a 0.635 mm thick Rogers RT6010 material (RT/duroid 6010 2011) with lateral dimensions of 24.65 mm by 17.5 mm. The ground plane is 11.3 mm by 17.5 mm. The input to the antenna is a 50 ohms microstrip. Metallization is with 0.05 mm thick copper. Adjustable parameters are $x = [v_1 \ v_2 \ v_3 \ w_1 \ w_2 \ w_3 \ w_4 \ u_1 \ u_2 \ u_3 \ u_4 \ u_5 \ u_6 \ u_7 \ u_8 \ u_9]^T = [8.9 \ 4.2 \ 3.0 \ 0.6 \ 1.2 \ 0.3 \ 0.3 \ 4.0 \ 1.5 \ 4.8 \ 1.8 \ 1.5 \ 4.0 \ 3.0 \ 3.35 \ 3.0]^T$ all in mm.

Antenna models are defined in the CST MWS environment and simulated with the CST MWS transient solver (CST Microwave Studio 2013). The high-fidelity model R_f is discretized with a mesh assuring that no noticeable change in the model

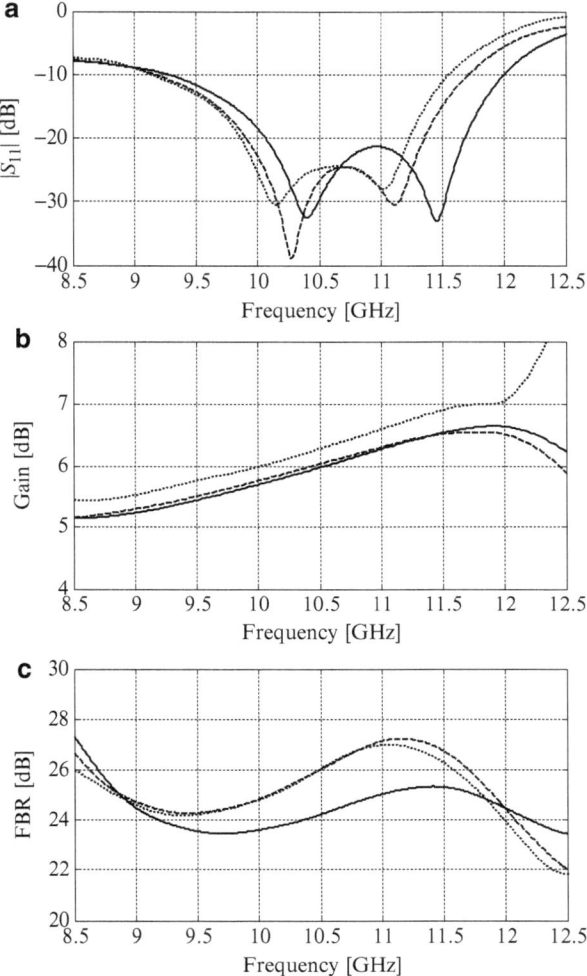

Fig. 5.6 Simulated responses of the planar Yagi antenna at a certain design: (**a**) reflection, (**b**) gain, and (**c**) front-to-back ratio (FBR). The high-fidelity model R_f (*solid line*), low-fidelity model R_{c1} (*dash line*), and low-fidelity model R_{c2} (*dot line*)

responses, both reflection and radiation, can be observed if the mesh would be made finer. To satisfy the above requirement, the number of mesh cells per wavelength at the center frequency of the simulation bandwidth has been found to be 45 at this particular design. As a result, the model R_f contains 1,611,624 mesh cells, and it is evaluated in 12 min 13 s with a 2.33 GHz 8 core CPU with 8 GB RAM computer. The low-fidelity model R_{c1} was made different from the model R_f only in mesh density and, therefore, in the number of mesh cells. In particular, the model R_{c1} is discretized for only 15 mesh cells per wavelength resulting in 96,000 mesh cells. R_{c1} simulation time is only 8 % of that of R_f. Responses of the models are shown in Fig. 5.6.

Another low-fidelity model R_{c2} has discretization density of the model R_{c1}. In addition to that other simplifications have been made as follows. Materials, Rogers RT60210 substrate and copper of metallization, have been made lossless. Metallization thickness has been set to 0. The number of PML layers of the absorbing boundaries has been changed from 6 to 4. The distances to the absorbing boundaries have been made 40 % of those of the models R_f and R_{c1}. Also, the termination condition has been set to −25 dB of the residual energy (CST Microwave Studio 2013) versus −40 dB of the models R_f and R_{c1}. The number of mesh cell in the model R_{c2} has been reduced not much compared to that of R_{c1}, to 90,000; however, its evaluation time is only 5 % of that of R_f.

By inspection of the reflection responses in Fig. 5.6a, one conclude that the low-fidelity models R_{c1} and R_{c2} have similar quality in overall when compared to the reflection response of the high-fidelity model R_f. Further, the gain responses of all models shown in Fig. 5.6b are quite close and consistent. The difference of the gain of the model R_{c2} from those of R_f and R_{c1} is due to its lossless description, and it can be easily taken into account up to 12 GHz, which is the high-end usable frequency of this design, e.g., through the space-mapping correction and/or adaptive response correction (Koziel et al 2008b, Koziel et al 2009b, and Koziel and Ogurtsov 2011a). The front-to-back ratio figures of the low-fidelity models are essentially the same and equally different from that figure of the model R_f.

From the surrogate-based optimization perspective, possible computational effort which would be required to correct the low-fidelity responses relative to the response of the high-fidelity model is expected to be about the same for both R_{c1} and R_{c2} (as both models are of similar quality); however, the use of model R_{c2} in the SBO process may result is a lower design cost (as R_{c1} is almost twice as fast as R_{c2}).

5.4   Need for Automated Selection of Model Fidelity

It is also worth to mention that a visual inspection of the model response and the relationship between the high- and low-fidelity models is an important part of the model selection process. It is essential that the low-fidelity model captures all important features of the high-fidelity model response. Nevertheless, because the low-fidelity model subjects two conflicting requirements of accuracy and speed (both depending, in general, on a particular location in the design space), the optimal choice is unclear a priori. In particular, the problem of model fidelity selection and handling should be quantitatively addressed within execution of a particular surrogate-based optimization algorithm. A more detailed discussion about the model fidelity selection and model fidelity effect on the performance of the surrogate-based optimization process, including both quality of the final design and the overall optimization, can be found in Chap. 12.

Chapter 6
Simulation-Based UWB Antenna Design

In this chapter, we illustrate applications of the SBO methodology to ultra-wideband (UWB) antennas. SBO techniques considered in this chapter include manifold mapping (MM), shape-preserving response prediction (SPRP), and adaptively adjusted design specification (AADS) techniques. Due to the behavior of reflection responses of UWB antennas (usually, relatively "flat" as compared to narrowband antennas), many other techniques might be suitable for this type of structures, cf. Sect. 6.4. The examples presented here illustrate that the computational costs of the SBO design process may be quite low and—in terms of the high-fidelity model evaluations—correspond to the number of design variables of the problem at hands.

6.1 UWB Monopole Matching with Manifold Mapping and Kriging

Consider an UWB monopole shown in Fig. 6.1 (Koziel and Ogurtsov 2012d). Design variables are $x = [h_0 \ w_0 \ a_0 \ s_0 \ h_1 \ w_1 \ l_{gnd} \ w_s]^T$. Other parameters are fixed: $l_s = 25$, $w_m = 1.25$, and $h_p = 0.75$, all in mm. The microstrip input of the monopole is fed through an edge mount SMA connector (AEP 2013). The initial design is $x^{(0)} = [18 \ 12 \ 2 \ 0 \ 5 \ 1 \ 15 \ 40]^T$ mm. Simulation time of the coarse-discretization model R_{cd} (~150,000 mesh cells) is 2 min, and that of the high-fidelity model R_f (~1,200,000 mesh cells) is 45 min (both at the initial design). Both models are evaluated using the transient solver of CST Microwave Studio (CST MWS 2013). The design specifications are $|S_{11}| \leq -10$ dB for 3.1 GHz to 10.6 GHz.

For this example the surrogate model was constructed with the manifold mapping (MM) (Echeverría and Hemker 2008) and the low-fidelity model R_{cd} (see also Sect. 4.5 of this book for a description of the MM technique). To reduce the design cost, the low-fidelity model was not used directly in the MM algorithm, but an auxiliary response surface approximation (RSA) model R_c was built using kriging (Kleijnen 2009, cf. Sect. 3.2.1) in the neighborhood of the approximate

S. Koziel and S. Ogurtsov, *Antenna Design by Simulation-Driven Optimization*, SpringerBriefs in Optimization, DOI 10.1007/978-3-319-04367-8_6, © Slawomir Koziel and Stanislav Ogurtsov 2014

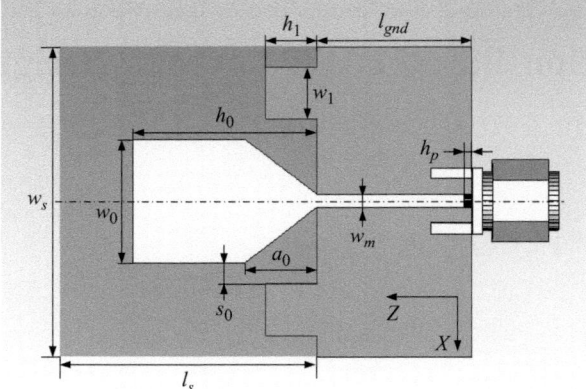

Fig. 6.1 UWB monopole, substrate shown semitransparent (Koziel and Ogurtsov 2012)

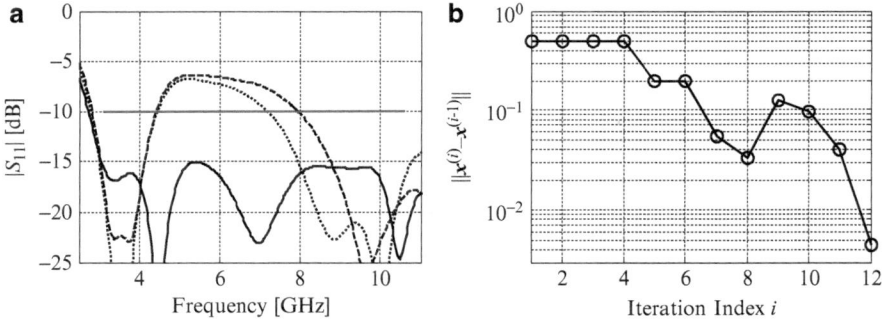

Fig. 6.2 UWB monopole: (**a**) R_f (*dashed line*) and R_{cd} (*dotted line*) models at the initial design, R_f (*solid line*) at the final design x^*, and (**b**) convergence of the MM algorithm (Koziel and Ogurtsov 2012)

optimum of model R_{cd} using 100 R_{cd} samples. Utilization of the RSA model allows us to eliminate the necessity of evaluating R_{cd} in the manifold mapping optimization process.

Optimization performed using the MM algorithm yielded the final design $x^* = [19.13\ 20.13\ 1.95\ 1.33\ 1.79\ 6.32\ 15.03\ 36.36]^T$ mm with $|S_{11}| < -15$ dB in the frequency band of interest. Figure 6.2a shows reflection responses of the high- and low-fidelity models at the initial design as well as the R_f response at the final design.

The MM algorithm used here had been enhanced with the low-fidelity model preconditioning by means of space mapping (Bandler et al. 2003) as well as the adaptive search radius scheme (Koziel et al. 2010a, b). This, together with the fact that toward the end of the MM optimization process (i.e., when $\|x^{(i)} - x^{(i-1)}\| \rightarrow 0$), the surrogate and the high-fidelity model Jacobians become more and more similar to each other, makes the MM algorithm capable for accurate location of the R_f optimum. The convergence plot for the MM algorithm is shown in Fig. 6.2b.

The total design cost is equivalent to about 21 high-fidelity model evaluations, and it comprises the following components: $100 \times R_{cd}$ (equivalent to about $4.4 \times R_f$) was spent to locate the optimum of R_{cd}; another 100 R_{cd} samples ($4.4 \times R_f$) were needed for the kriging model in the neighborhood of the R_{cd} optimum; finally, 12 evaluations of model R_f were performed through the algorithm run.

6.2 UWB Dipole

Consider a planar dipole antenna shown in Fig. 6.3 comprising a planar dipole as the main radiator element and two additional strips. The substrate is Rogers RT5880. Predefined dimensions are $a_1 = 0.5$ mm, $w_1 = 0.5$ mm, $l_s = 50$ mm, $w_s = 40$ mm, and $h = 1.58$ mm. The design variables are $\mathbf{x} = [l_0 \ w_0 \ a_0 \ l_p \ w_p \ s_0]^T$. The initial design is $\mathbf{x}^{init} = [20 \ 10 \ 1 \ 10 \ 8 \ 2]^T$ mm. The design objective is to obtain $|S_{11}| \leq -12$ dB for 3.1 GHz to 10.6 GHz. The high-fidelity model R_f of the antenna structure, having 10,250,412 mesh cells at the initial design $\mathbf{x}^{init}$, is evaluated in 44 min using the CST MWS transient solver. The low-fidelity model R_{cd}, having 108,732 cells at $\mathbf{x}^{init}$, is evaluated with the same solver in 43 s.

The shape-preserving response prediction (SPRP) technique was applied here to adjust the dipole dimensions. A detailed description of SPRP can be found in Sect. 4.3 of this book. As the first step, the approximate optimum of R_{cd}, $\mathbf{x}^{(0)} = [18.66 \ 12.98 \ 0.526 \ 13.717 \ 8.00 \ 1.094]^T$ mm, was found. The computational cost of this step is 127 evaluations of R_{cd}; it corresponds to about two evaluations of R_f. Figure 6.4a shows the reflection responses of R_f at both $\mathbf{x}^{init}$ and $\mathbf{x}^{(0)}$, as well as the response of R_{cd} at $\mathbf{x}^{(0)}$. Then the final design $\mathbf{x}^{(2)} = [19.06 \ 12.98 \ 0.426 \ 13.52 \ 6.80 \ 1.094]^T$ mm with $|S_{11}| \leq -13.5$ dB for 3.1 GHz to 10.6 GHz had been obtained in two iterations of the SPRP algorithm. Figures 6.4b and 6.5 show the responses of the final design. The design cost is detailed in Table 6.1.

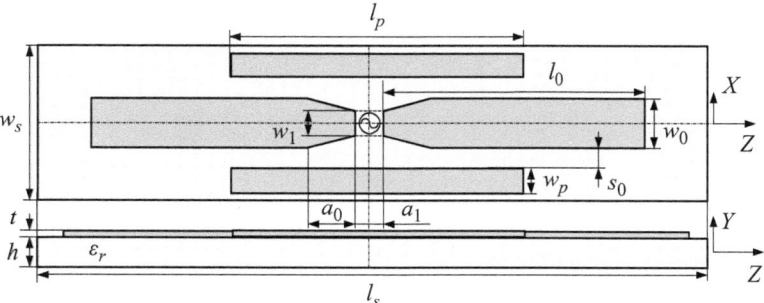

Fig. 6.3 UWB dipole antenna geometry: top and side views. The *dash-dot lines* show the electric (YOZ) and the magnetic (XOY) symmetry walls (Koziel and Ogurtsov 2011e)

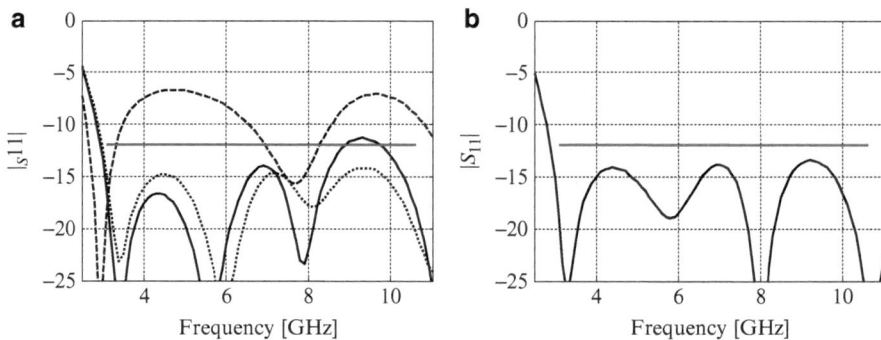

Fig. 6.4 UWB dipole reflection response: (**a**) high-fidelity model (*dashed line*) at the initial design x^{init} and high- (*solid line*) and low-fidelity (*dotted line*) model at the approximate low-fidelity model optimum $x^{(0)}$ and (**b**) high-fidelity model at the final design (Koziel and Ogurtsov 2011e)

Fig. 6.5 UWB dipole at the final design: IEEE gain pattern (×-pol.) in the XOY plane at 4 GHz (*thick solid*), 6 GHz (*dash-dot line*), 8 GHz (*dash line*), and 10 GHz (*solid line*) (Koziel and Ogurtsov 2011e)

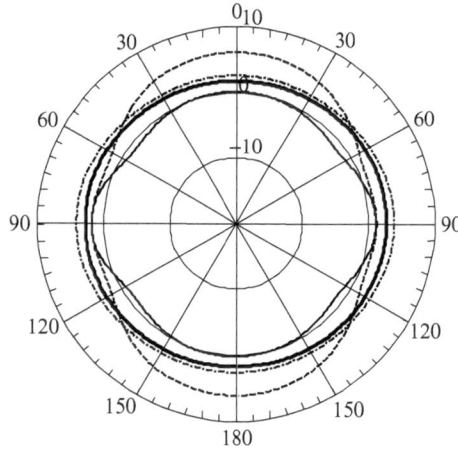

Table 6.1 UWB Dipole: optimization cost (Koziel and Ogurtsov 2011e)

Algorithm component	Number of model evaluations	Evaluation time	
		Absolute (min)	Relative to R_f
Evaluation of R_{cd}[a]	$233 \times R_{cd}$	167	3.8
Evaluation of R_f[b]	$3 \times R_f$	132	3.0
Total optimization time	N/A	299	**6.8**

[a]Includes initial optimization of R_{cd} and optimization of SPRP surrogate
[b]Excludes evaluation of R_f at the initial design

6.3   UWB Vivaldi Antenna

Consider a Vivaldi antenna (Qing and Chen 2004) shown in Fig. 6.6. Design variables are $x = [a_1 \, a_2 \, b_1 \, b_3 \, h_1 \, h_2 \, d_1]^T$. The profile of the antipodal metal fins is with arcs of ellipses; for the upper fin they are *BC*, *DE*, and *DB*. The point *A* is the center

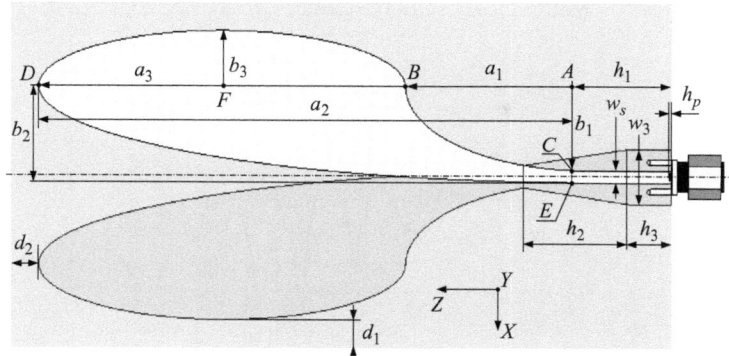

Fig. 6.6 UWB Vivaldi antenna, substrate shown transparent (Ogurtsov and Koziel 2011a)

of two ellipses with the arks of BC and DE and the semiaxes of a_1 and b_1 and a_2 and b_2, respectively. The point F is the center of the ellipse with the semiaxes of a_3 and b_3. Here $a_3 = (a_2 - a_1)/2$, $b_2 = b_1 + w_s$, and $d_2 = d_1$. Other parameters are fixed: $w_s = 2.15$, $w_1 = 12.9$, $h_3 = 5$, and $h_p = 0.5$ (all mm). Substrate is 0.787 mm thick Rogers RT5880. Metallization is with 0.05 mm thick copper. An edge mount SMA connector (AEP 2013) interfaces the antenna with the 50 Ω coaxial. And a pair of through vias connects the connector tips to the microstrip ground. The initial design is $x^{(0)} = [30\ 50\ 10\ 10\ 100\ 20\ 2]^T$ mm.

At the initial design, the fine model R_f (9,851,880 mesh cells) and the coarse-discretization model R_{cd} (1,101,120 mesh cells) are evaluated in 1 h 45 min and 6 min, respectively, using the transient solver of CST MWS (CST MWS 2013). The design requirements are $|S_{11}| \leq -10$ dB for 3.1 GHz to 10.6 GHz; the antenna is to be of high-gain and end-fire radiation over the bandwidth.

The antenna was optimized using the adaptively adjusted design specification (AADS) technique introduced in Koziel (2010b, c). AADS consists of the two steps which are iterated if necessary:

1. Modify the original design specifications in order to take into account the difference between the responses of R_f and R_{cd} at their characteristic points.
2. Obtain a new design by optimizing the low-fidelity model R_{cd} with respect to the modified specifications.

Notice that in the AADS approach, the optimizer works on the model R_{cd} directly so that no surrogate similar to those of the previous examples of this chapter should be created. In a way, the modified and updated design specifications (imposed on $|S_{11}|$ in this particular example) play a role of the AADS surrogate model. Detailed formulation and a discussion of the AADS optimization technique can be found in Sect. 4.6 of this book.

Two iterations of the AADS technique yielded the final design $x^{(2)} = [37.07\ 33.35\ 24.75\ 53.34\ 123.05\ 32.81\ 1.23]^T$ mm which has $|S_{11}| < -15.7$ dB for the 3.1 GHz to 10.6 GHz bandwidth. The reflection response of the high- and low-fidelity model at

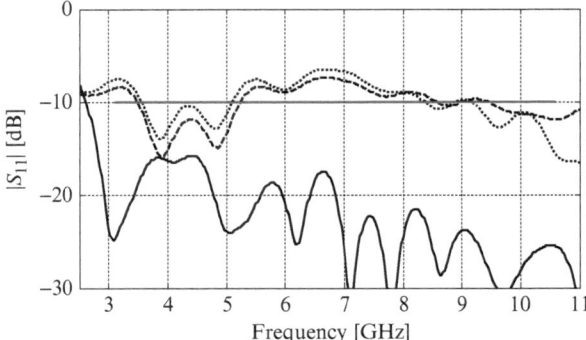

Fig. 6.7 UWB Vivaldi antenna: the high-, R_f (*dash line*), and the low-fidelity, R_{cd} (*dotted line*), model responses at the initial design $x^{(0)}$, as well as the high-fidelity model R_f (*solid line*) at the final design (Ogurtsov and Koziel 2011a)

Fig. 6.8 Gain [dBi] (x-pol.) of the Vivaldi antenna: pattern cut in the *YOZ* plane at 4 GHz (*solid line*), 6 GHz (*dash line*), 8 GHz (*dash-dot line*), and 10 GHz (*solid dots*). 90° on the left, 0°, and 90° on the right are for *Y*, *Z*, and −*Y* directions, respectively (Ogurtsov and Koziel 2011a)

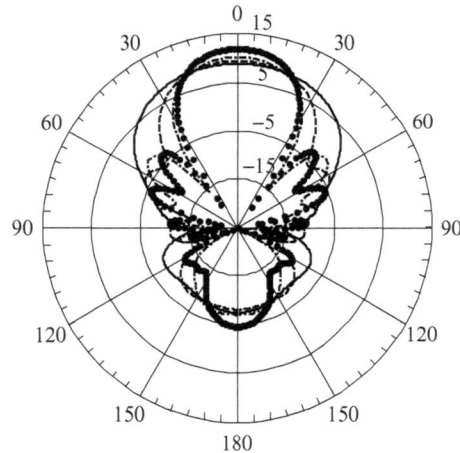

Table 6.2 UWB Vivaldi antenna: optimization costs (Ogurtsov and Koziel 2011a)

Algorithm component	Model evaluations	CPU time	
		Absolute	Relative to R_f
Optimizing R_c	$195 \times R_{cd}$	19 h 30 min	11.1
Evaluation of R_f	$2 \times R_f$	3 h 30 min	2.0
Total cost[a]	N/A	23 h	**13.1**

[a]Excludes R_f evaluation at the initial design

the initial design as well as the R_f response at the final design is shown in Fig. 6.7. Gain patterns of the final design at selected frequencies are shown in Fig. 6.8.

The total number of evaluations of R_{cd} in the optimization process is 195. Table 6.2 shows the computational cost of the optimization: the total optimization time corresponds to about 13 evaluations of the high-fidelity model.

6.4 Discussion

Only three specific SBO techniques have been illustrated in this chapter. They can be considered as representative methods that can be used to handle UWB antenna reflection responses. A number of other SBO techniques have also been applied to UWB antennas. These SBO techniques include SBO with Bayesian support vector regression models (Jacobs et al. 2013), SBO with co-kriging-based surrogates (Koziel et al. 2013), SBO with SM corrected Cauchy-approximation surrogates (Koziel et al. 2011e, Koziel et al. 2012a, b, c, d, e), SM with kriging-based surrogates (Ogurtsov and Koziel 2010), SM combined with adjoint sensitivities (Koziel and Ogurtsov 2012a), and variable-fidelity optimization algorithm (Koziel and Ogurtsov 2011f).

It should be emphasized that ultra-wideband antenna responses can be handled using a variety of methods because they are relatively "flat" as compared to, e.g., responses of narrowband antennas or filters. For the same reason, UWB surrogates can often be created using very simple techniques such as additive or multiplicative response correction and frequency scaling (Koziel and Ogurtsov 2012a, b, c).

Chapter 7
Optimization of Dielectric Resonator Antennas

In this chapter we demonstrate the use of SBO methodology for design optimization of dielectric resonator antennas (DRAs). Analysis, design verification, and optimization of DRAs are all simulation based because analytical models of DRA (Petosa 2007), though indispensable, can be used mostly to estimate initial designs which should be further tuned to account for particular installation environment, housing, and feeding (Kishk and Antar 2007). Simulation-based optimization of DRAs is normally associated with multiple evaluations of their discrete full-wave models, which turn to be computationally expensive, even for a single accurate simulation due to the high-permittivity dielectric resonator core and full-wave nature of the underlying physics. SBO techniques alleviate this problem substantially by shifting the computational burden to the DRA surrogate so that optimization of realistic DRA models can be conducted.

7.1 DRA with a Substrate-Integrated Cavity

In this section, we describe an optimization of a 2.4–2.5 GHz DRA coupled to a substrate-integrated cavity. The DRA is directly fed by a grounded coplanar waveguide (GCPW) through two slots in the upper ground plane. Additionally, to reduce the noise emitted into the substrate, a substrate-integrated resonator cavity is introduced underneath the dielectric resonator antenna. SBO methodology exploiting a fast surrogate DRA model is used to adjust the dimensions of the DRA as well as to reduce the overall design time.

Consider a DRA shown in Fig. 7.1 (Ogurtsov and Koziel 2011b). It comprises two slot-fed coupled rectangular dielectric resonators (DRs) (Deng et al. 2004) installed above a layer with upper and lower metal grounds. The DRs are covered by a polycarbonate housing which has dielectric constant and loss tangent of 2.7 and 0.01, respectively. The housing is mounted on the board with four through M2 bolts. Feeding of the DRA is with a $50\,\Omega$ GCPW terminated by two symmetrical slots (width s_1 and length x_1) shown in Fig. 7.2a and exiting two $\mathrm{TE}^x_{\delta 11}$ DRs.

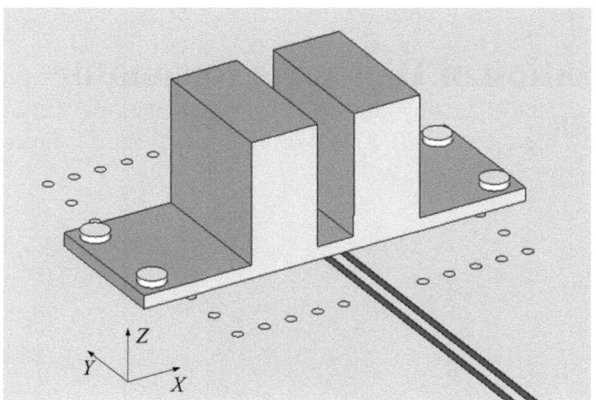

Fig. 7.1 DRA 3D view (Ogurtsov and Koziel 2011b)

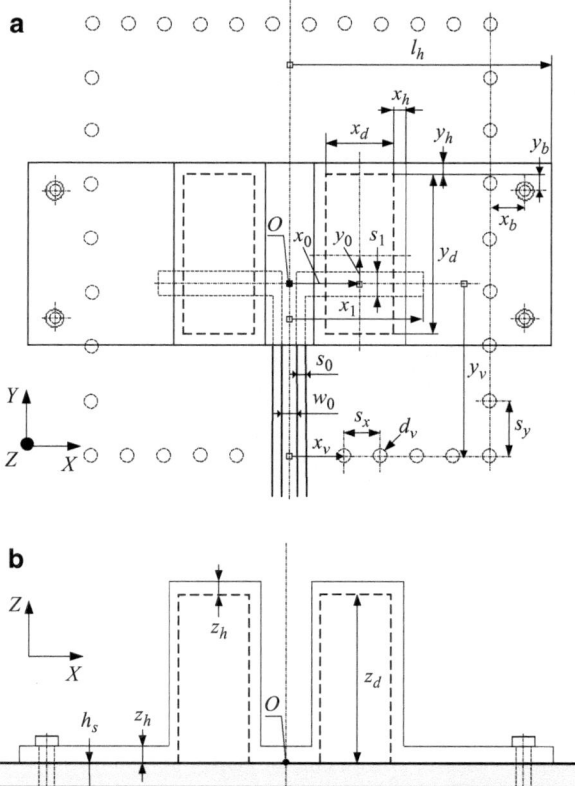

Fig. 7.2 DRA: (**a**) top view; (**b**) front view (vias not shown) (Ogurtsov and Koziel 2011b)

Figures 7.1 and 7.2a also show vias forming a substrate-integrated (cage-like) cavity. The relative permittivity and loss tangent of the DR ceramic cores are 36 and 1e-4, respectively.

While the expected resonant frequency and unloaded Q-factor of an isolated single DR working on $TE^x_{\delta 11}$ mode can be easily estimated (Petosa 2007), the effect of coupling between the DR cores, presence of the feeding slots, and housing require full-wave simulations to describe both the reflection and radiation response of the DRA.

The lower ground provides isolation of the DRA from the layers underneath the considered structure; however, parallel plate modes can be launched in the substrate by the feeding slots. This undesirable phenomenon results in the increase of the substrate noise as well as degradation of the antenna gain. Undesirable emission of the signal in the substrate can be suppressed with through-vias connecting the upper and lower grounds of the layer and, therefore, forming a substrate-integrated cavity underneath the DRA. This modification being straightforward as a concept introduces additional degrees of freedom to the design. As a result, simulation-driven design of the DRA becomes hardly feasible through parameter sweeps. To achieve the design goals, we adopt an automated SBO procedure of Koziel (2009). We use CST MWS to define the DRA model and evaluate its response for different combinations of design parameters.

There are 11 design variables: $\mathbf{x} = [x_0\ y_0\ x_d\ y_d\ z_d\ s_1\ x_1\ x_v\ y_v\ s_x\ s_y]^T$, where x_0 and y_0 are location of the center of a DR relative to the origin of the coordinate system marked by O in Fig. 7.2; x_d, y_d, and z_d are dimensions of the DR ceramic cores; s_1 and x_1 are dimensions of the DR energizing slots; and x_v, y_v, s_x, and s_y describe via locations and in row spacing as shown in Fig. 7.2a. The substrate-integrated cavity is defined with 10 vias in the lower (horizontal) row, 11 vias in the upper (horizontal) row, and 9 vias in the vertical rows; see Fig. 7.2a. Other dimensional parameters are fixed as follows. Substrate is 2.5 mm thick RT6010. Dimensions of the input GCPW are signal trace width, w_0, of 1.5 mm and spacing, s_0, of 1 mm. Diameter of the vias, d_v, is 1.5 mm. Thicknesses of the polycarbonate housing, x_h, y_h, and z_h, are 2 mm.

Location of the mounting bolts are described by $x_h = s_x$ and $y_h = 1$ mm. The bolt heads are 4 mm in diameters and 1 mm thick. Lateral extension of the housing is $l_h = x_v + 5s_x + 3$ [mm]. The entire structure has a magnetic symmetry plane which is shown with vertical dash-dot lines in Fig. 7.2. Metallization of the ground and GCPW trace is with 1.5 oz (0.05 mm thick) copper.

Design requirements are the following: reflection coefficient, $|S_{11}|$, should be lower than -20 dB, and the gain is to be higher than 3dBi for $\theta = 0^0$ (Z-direction), both over the for the 2.4–2.5 GHz.

The high-fidelity model $\mathbf{R}_f$ is CPU intensive, about 1 h per design. Therefore, we use a coarse-discretization EM model (denoted as $\mathbf{R}_{cd}$) to obtain an approximation of the design which is further refined using kriging and the conventional SM algorithm (Koziel 2009). The design optimization procedure is the following (see also Sect. 4.2.4 of this book for more details):

1. Starting from $\mathbf{x}^{init}$, find an approximate optimum $\mathbf{x}^{(0)}$ of the coarse-discretization model $\mathbf{R}_{cd}$. Here, we use a pattern search algorithm (Kolda et al. 2003).

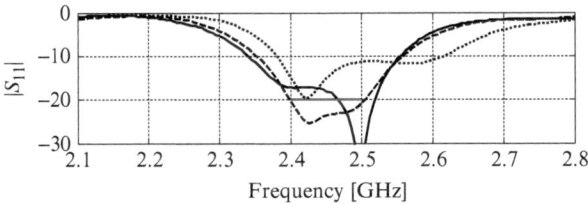

Fig. 7.3 Design procedure (first stage): $|S_{11}|$ of the coarse-discretization DRA model at the initial design (*dotted line*), $|S_{11}|$ of the coarse-discretization model at its optimized design (*dashed line*), and $|S_{11}|$ of the high-fidelity model at the coarse-discretization model optimum (*solid line*). Specifications are shown using horizontal line (Ogurtsov and Koziel 2011b)

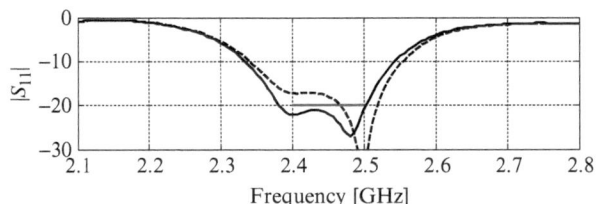

Fig. 7.4 Design procedure (second stage): $|S_{11}|$ response of the high-fidelity model at the coarse-discretization model optimum (*dashed line*) and at the final design obtained using space-mapping optimization with kriging coarse model (*solid line*). Design specifications shown with the *horizontal line* (Ogurtsov and Koziel 2011b)

2. Sample R_{cd} in the neighborhood of $x^{(0)}$ and construct a response surface approximation model R_c (here, kriging as an approximation technique).
3. Find a high-fidelity model optimum by applying the SM algorithm (Koziel et al. 2009c) with R_c as an underlying coarse model.

The operation of the design procedure is illustrated in Figs. 7.3 and 7.4. The SM algorithm yields approximate solutions to the original problem $x^* = \text{argmin}\{x: U(R_f(x))\}$, where U is an objective function that measures the violation of the design specifications (here, 20 dB minus the maximum of $|S_{11}|$ in the frequency band of interest). At iteration i a new design $x^{(i)}$ is generated so that $x^{(i)} = \text{argmin}\{x: U(R_s^{(i)}(x))\}$, where $R_s^{(i)}$ is the surrogate model, defined here as $R_s^{(i)}(x) = R_c(x + c^{(i)}) + d^{(i)}$. The vector $c^{(i)}$ is obtained in the parameter extraction process (Bandler et al. 2003) to minimize $\|R_f(x) - R_c(x + c)\|$. The vector $d^{(i)} = R_f(x) - R_c(x + c^{(i)})$. The surrogate constructed by means of coarse-discretization model data and the SM alignment allows us to locate the high-fidelity model optimum in a few iterations. Step 2 is necessary because the SM algorithm requires a large number of coarse model evaluations so that a direct use of the coarse-discretization model R_{cd} is not practical. The design procedure described here is implemented and executed using the SMF optimization environment (Koziel and Bandler 2007a). Starting from $x^{\text{init}} = [7.75\ 5\ 6\ 16.5\ 18\ 2\ 10.75\ 6\ 14\ 4\ 6]^T$ mm, the final design was found to be $x^* = [7.62\ 5.70\ 6.2\ 16.43\ 17.9\ 1.9\ 10.45\ 6.08\ 13.83\ 4.37\ 6.03]^T$ mm. The design response meets the

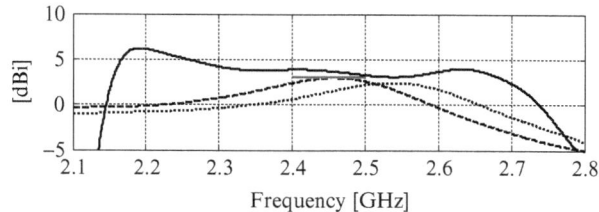

Fig. 7.5 DRA, IEEE gain response in Z-direction at the final design: with substrate-integrated cavity, x^* (*solid line*); no vias, $x^{*,n.v}$ (*dashed line*); and no vias, $x^{**,n.v.}$ (*dotted line*). Design specifications shown with the horizontal line (Ogurtsov and Koziel 2011b)

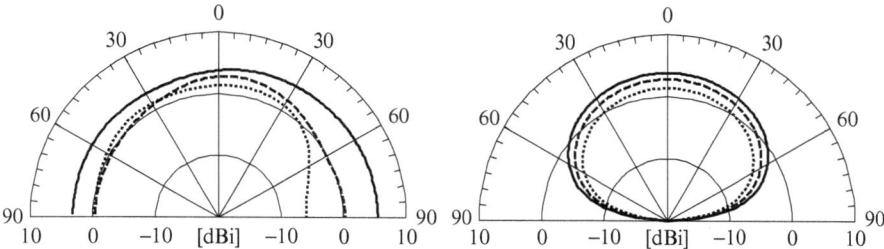

Fig. 7.6 DRA, IEEE gain at 2.45 GHz: (*left*) co-pol. in the *E*-plane (*YOZ*), the right sector is for the positive *Y*-direction; (*right*) x-pol. in the *H*-plane (*XOZ*). Design with substrate-integrated cavity, x^* (*solid line*); designs without vias, $x^{*,n.v}$ (*dashed line*) and $x^{**,n.v.}$ (*dotted line*) (Ogurtsov and Koziel 2011b)

Fig. 7.7 DRA, $|S_{11}|$ response at the final design: with substrate-integrated cavity, x^* (*solid line*); no vias, $x^{*,n.v}$ (*dashed line*); and no vias, $x^{**,n.v.}$ (*dotted line*) (Ogurtsov and Koziel 2011b)

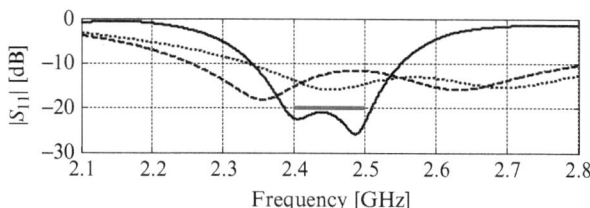

specifications; its $|S_{11}|$ is shown in Fig. 7.4, the gain versus frequency for $\theta = 0^0$ is shown in Fig. 7.5, and the gain pattern cuts at 2.45 GHz are shown in Fig. 7.6. For comparison, the DRA without substrate-integrated cavity was also considered. In this case there were seven design variables $x^{*,n.v.} = [x_0 \ y_0 \ x_d \ y_d \ z_d \ s_1 \ x_1]^T$. The optimal designs found for this case (no cavity) do not satisfy the design requirements ($|S_{11}| < -20$ dB, gain ($\theta = 0^0$) > 3dBi, for 2.4–2.5 GHz). Figures 7.5, 7.6, and 7.7 show responses of the two alternative designs, $x^{*,n.v} = [7.65 \ 5.51 \ 5.39 \ 16.20 \ 19.45 \ 0.263 \ 10.05]^T$mm ($|S_{11}| < -11.5$ dB, gain ($\theta = 0^0$) > 2.5dBi) and $x^{***,n.v} = [6.79 \ 5.25 \ 5.68 \ 16.22 \ 19.97 \ 0.250 \ 9.46]^T$mm ($|S_{11}| < -13.5$ dB, gain ($\theta = 0^0$) > 0.5dBi).

7.2 Suspended Brick DRA

Consider a DRA shown in Fig. 7.8 (Koziel and Ogurtsov 2011h). The rectangular dielectric resonator (DR) is excited at the $TE\delta_{11}$ mode with a $50\,\Omega$ microstrip through a slot made in the metal ground plane. Substrate is 0.5 mm thick RO4003C material of infinite lateral extends. Metallization of the ground and the microstrip trace (the width w_0 of 1.15 mm) is with 0.05 mm thick copper. DR relative permittivity and loss tangent are 10 and 1e-4, respectively.

The design task is to adjust dimensions of the DR brick, a_x, a_y, and a_z; the slot dimensions, u_s w_s; the length of the microstrip slab, y_s; and location of the DR relative to the slot, a_c, so that the DRA bandwidth is to be 5.1GHz to 5.9 GHz with the -15 dB of reflection level. In addition also the back radiation (down the substrate) should be kept as low as possible. A simplest concept to enhance impedance bandwidth of a DRA is to suspend it above the ground plane (Petosa 2007, Kishk and Antar 2007). Therefore, the DR core is placed on Teflon bricks, as shown in Fig. 7.8b, c, bringing additional degrees of freedom g_1 and b_y into the design. As a result, the design variables are $[a_x\ a_y\ a_z\ a_c\ u_s\ w_s\ y_s\ g_1\ b_y]^T$. Additionally, we consider polycarbonate housing with relative permittivity of 2.8 and loss tangent of 0.01. The fixed dimensions of the housing (see Fig. 7.8 b, c) are $d_x=d_y=d_z=1$ mm, $d_{zb}=2$ mm, $b_x=2$ mm, and $c_x=6.5$ mm. The initial design is $x^{(0)}=[8.0\ 14.0\ 9.0\ 0.0\ 1.75\ 10.0\ 3.0\ 1.5\ 6.0]^T$ mm.

The high-fidelity model R_f, 763,840 mesh cells at the initial design $x^{(0)}$, was simulated using the CST MWS transient solver (CST MWS 2013) for 18 min 22 s. The design objective was to obtain $|S_{11}|\leq-15$ dB for 5.1 GHz to 5.9 GHz. Requirements imposed on realized gain were the following: it should be at least 4 dB the zero zenith angle and realized gain of back radiation should be less than -10 dB. These constrains were imposed over the impedance bandwidth. In this case, we exploited the low-fidelity model R_{cd}, also evaluated with CST MWS (30,720 mesh cells at $x^{(0)}$, evaluation time of 44 s).

As indicated in Fig. 7.9, discrepancy between reflection responses of the low- and high-fidelity models was substantial. In order to accommodate this discrepancy

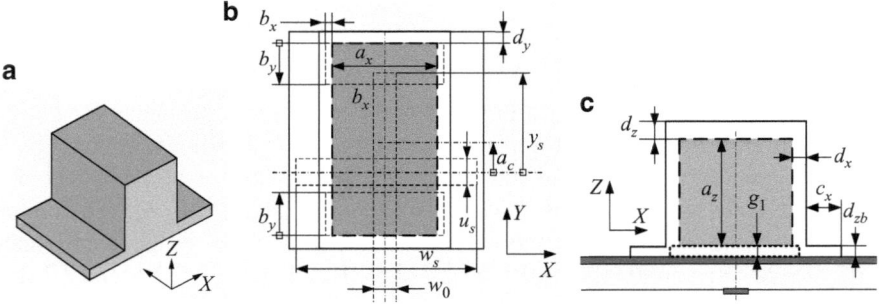

Fig. 7.8 Suspended DRA: (**a**) 3D view of its housing, top (**b**) and front (**c**) views (Koziel and Ogurtsov 2011h)

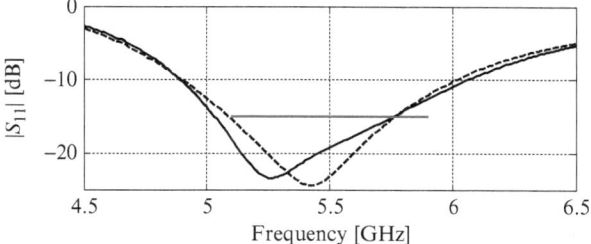

Fig. 7.9 DRA reflection response at the initial design $x^{(0)}$ with the high-fidelity model R_f (*solid line*) and low-fidelity model R_{cd} (*dashed line*) (Koziel and Ogurtsov 2011h)

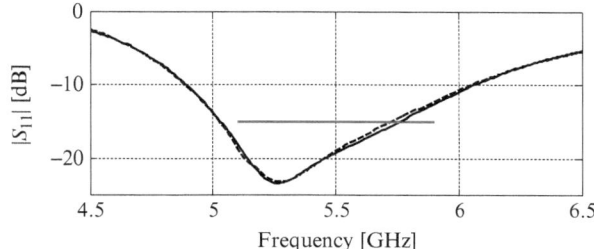

Fig. 7.10 DRA reflection response at the initial design $x^{(0)}$ with the high-fidelity model R_f (*solid line*) and SM-corrected kriging model R_s (*dashed line*) (Koziel and Ogurtsov 2011h)

and still keep the design cost low, we adopted the strategy described in Sect. 4.2.4. First, a kriging-interpolation-based (Queipo, *et al.* 2005) surrogate R_c of the low-fidelity model had been established in the vicinity of $x^{(0)}$ defined as $[x^{(0)} - \delta, x^{(0)} + \delta]$, where $\delta = [0.5\ 0.5\ 0.5\ 0.5\ 0.5\ 0.25\ 0.25\ 0.25\ 0.5]^T$ mm. To set up the surrogate, we used 200 samples allocated using Latin hypercube sampling (Beachkofski and Grandhi 2002).

Next, the kriging surrogate served as a coarse model for the space-mapping (SM) algorithm (Koziel et al. 2008b) which was the optimization engine. The SM model was of the form $R_s(x) = R_c(x + c)$, where c was a vector obtained using the parameter extraction process that aimed at minimizing $\|R_f(x) - R_c(x + c)\|$. The SM-corrected R_c had been used to yield an approximated optimum of the high-fidelity model. This kind of correction was sufficient as indicated in Fig. 7.10 showing a good agreement at $x^{(0)}$ between the high-fidelity model and SM-corrected kriging surrogate model. Since vector c was design dependent so that a few SM iterations were necessary to yield an optimized design of the high-fidelity model.

The use of the kriging model of R_{cd} instead of R_{cd} itself in the SM algorithm directly was necessary because the parameter extraction process required a substantial amount of model evaluations, which, together with the additional cost of optimizing the surrogate, would degrade the computational efficiency of the design process. Instead, a fixed number of evaluations (here, 200) had been executed initially with no further cost associated to surrogate model evaluation subsequently.

The final design $x^{(2)} = [7.675\ 13.9\ 8.875\ 0\ 1.95\ 10.0\ 2.825\ 1.6\ 5.9]^T$ mm had been obtained in two iterations of the SM algorithm with $|S_{11}| < -16$ dB for 5.1 GHz to 5.9 GHz as shown in Fig. 7.11. Figure 7.12 shows the realized gain response at the final design for selected zenith angles in the E-plane. The optimization costs are summarized in Table 7.1.

Fig. 7.11 Dielectric
resonator antenna: high-
fidelity model response at the
final design $x^{(2)}$ obtained after
two space-mapping iterations
(Koziel and Ogurtsov 2011h)

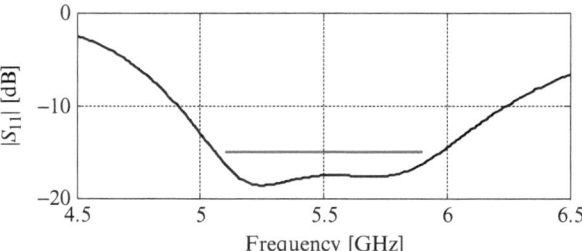

Fig. 7.12 Realized gain of
the DRA at the final design in
the E-plane (YOZ): zenith
angle of 0° (*thick solid line*);
back radiation, zenith angles
of 135° (positive *Y*-direction,
thin solid line), 180° (*dash
line*), and 135° (negative
Y-direction *dash-dot line*).
Design constrains are shown
with the *horizontal lines* at
the 3 dB and −10 dB levels
(Koziel and Ogurtsov 2011h)

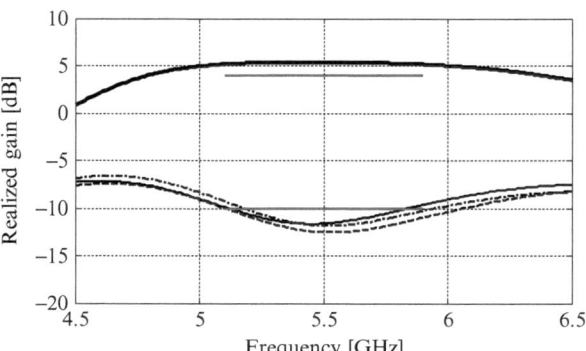

Table 7.1 Suspended DRA with housing: optimization cost (Koziel and Ogurtsov 2011h)

Algorithm component	Model evaluations	Evaluation time	
		Absolute (min)	Relative to R_f
Evaluation of R_{cd}[a]	$200 \times R_{cd}$	146.7	8.0
Evaluation of R_f[b]	$2 \times R_f$	36.7	2.0
Total optimization time	N/A	183.4	10.0

[a]Includes evaluations of R_{cd} used to set up the kriging model
[b]Excludes evaluation of R_f at the initial design

7.3 Optimization of DRA for Two Installation Scenarios

Design of a DRA for matched operation in two different installation environments
over the 4–6 GHz band is demonstrated below as our final example (Ogurtsov and
Koziel 2011c). In the first scenario, the $TM_{01}\delta$ DRA is installed at the infinite
metal ground (Fig. 7.13a), while in the second scenario, this DRA has a finite
circular ground and radiates into free space (Fig. 7.13b). Problem like that, i.e.,
meeting design specifications in different installation environments, can be often
encountered by an antenna designer not only for DRAs. A solution cannot be
quickly delivered with a parametric study. On the other hand, this problem can be

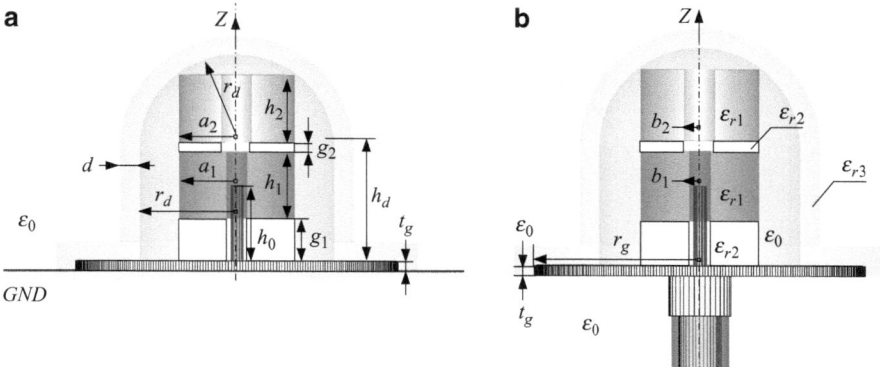

Fig. 7.13 DRA side views: (**a**) the DRA installed at the infinite metal ground; (**b**) the same DRA with the finite ground only. A feeding cable with 1 mm polyimide coating is shown on (**b**). The dome and DRA rings are shown semitransparent (Ogurtsov and Koziel 2011c)

solved with the SBO approach which combines coarse-discretization models of the DRA and the AADS technique (Koziel 2010b; see also Sect. 4.6 of this book). AADS can efficiently handle coarse-discretization models which are relatively expensive, in particular, not as computationally cheap as circuit equivalents (Bandler et al. 2004a, b).

A rotationally symmetric DRA (Shum and Luk 1995) is shown in Fig. 7.13. It comprises two $TM_{01}\delta$ dielectric resonator (DR) rings with relative permittivity, ε_{r1}, of 36, two supporting Teflon rings, Teflon filling, and finite ground ($t_g = 1$ mm). Teflon permittivity, ε_{r2}, is 2.08. The DRA is covered by a polycarbonate ($\varepsilon_{r3} = 2.7$) dome. Thickness of the dome shell, d, is 2 mm. Dielectric loss tangents are 10e-4 for the DRs, 4e-4 for Teflon, and 1e-2 for the dome, all at 6 GHz. The radii of the supporting rings are set to be equal to the radii of the DR above them. All metal parts have conductivity of copper. The inner conductor of the 50 Ω coax is extended in the DRA h_0 above the ground as a probe with 1.27 mm in diameter. The coax is also filled by Teflon.

Design variables are inner and outer radii of the DRs, heights of the DRs and the supporting rings, the probe length, dome height and radius, and radius of the DRA ground, namely, $\boldsymbol{x} = [a_1\ a_2\ b_1\ b_2\ h_1\ h_2\ g_1\ g_2\ h_0\ h_d\ r_d\ r_g]^T$. The design objective is $|S_{11}| \leq -15$ dB in the frequency band 4 GHz to 6 GHz for the DRA that can be installed in two environments shown in Fig. 7.13. The following reasons make this problem challenging: (1) the large number of variables for a simulation-based design, namely, 12; (2) high computational cost of a single simulation run; and (3) two installation environments considered at the same time. The last issue not only increases the computational cost (two EM analyses have to be performed in each step of the optimization process, one for each environment) but also requires finding a trade-off between optimal designs of each environment.

The problem is formulated here as a nonlinear minimization with the minimax objective function of the form $H(\boldsymbol{x}) = \max\{H_1(\boldsymbol{x}), H_2(\boldsymbol{x})\}$, where $H_i(\boldsymbol{x}) = \max\{|S_{11}(\boldsymbol{x},f)|$

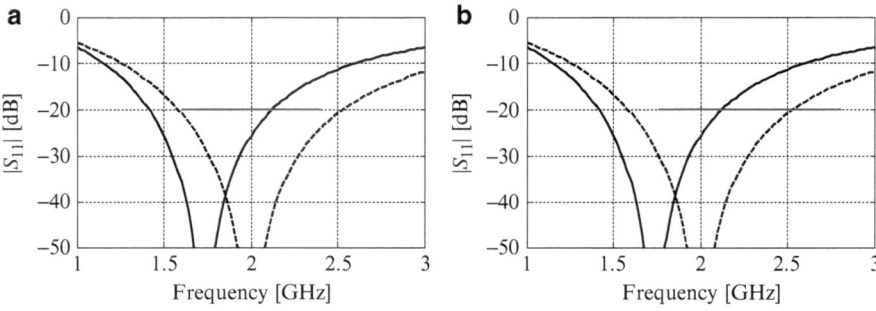

Fig. 7.14 AADS concept: (**a**) high- (*solid line*) and low-fidelity (*dashed line*) model responses as well as original design specifications; (**b**) modified specifications accounting for the discrepancy between the models

: 4 GHz$\leq f \leq$6 GHz}, $i=1$, 2, is the maximum reflection in the frequency band of interest for the first and second installation scenario, respectively. The design specifications are satisfied if $H(x) \leq -15$ dB. Note that each evaluation of the objective function requires two high-fidelity EM simulations. The EM models of the DRA are simulated with the MWS transient solver (CST MWS 2013).

The initial approximation of the optimal design, x^{init}, is obtained by optimizing the coarse-discretization EM antenna models using the pattern search algorithm (Kolda et al. 2003). These coarse models are faster than high-fidelity ones (about 15 times); however, they are also less accurate so that the discrepancy in $|S_{11}|$ between low- and high-fidelity models depends on frequency and can be as large as 5–10 dB.

Next, the design is further improved using the AADS technique so that the level of satisfying/violating the modified specifications by the low-fidelity model response corresponds to the satisfaction/violation levels of the original specifications by the high-fidelity model response. Figure 7.14 explains the AADA concept assuming that the low-fidelity response is shifted by Δf to higher frequencies compared to that of the high-fidelity model and that this shift is the only discrepancy between the models. More detailed formulation of AADS can be found in Sect. 4.6 of this book.

Design optimization continues from $x^{init} = [a_1\, a_2\, b_1\, b_2\, h_1\, h_2\, g_1\, g_2\, h_0\, h_d\, r_d\, r_g]^T = [6.9$ 6.9 1.05 1.05 6.2 6.2 2.0 2.0 6.8 12.0 10 16.5]^T mm which is far from meeting the design requirements (see Fig. 7.15a). At x^{in} the high-fidelity model with the finite ground (Fig. 7.13b) has 4,369,634 mesh cells and that with the infinite ground (Fig. 7.13a) has 4,006,017 mesh cells; their run times are 10,088 s and 8,697 s, respectively. The coarse-discretization model with the finite ground has 696,135 mesh cells and that with the infinite ground has 600,848 mesh cells; their run times are 684 s and 577 s, respectively.

The final design, $x^* = [5.9\, 1.05\, 7.825\, 5.9\, 1.8\, 7.95\, 4.75\, 0.90\, 7.75\, 13.50\, 10.0$ 18.40]^T mm, is obtained using the AADS technique. Figure 7.15b shows the DRA reflection responses at the final design. The radiation responses of the final design at selected frequencies are shown in Fig. 7.16. The total design cost being equivalent to about 20 high-fidelity model evaluations shows that our optimization procedure is quite efficient for the 12 design variables.

Fig. 7.15 $|S_{11}|$ of the initial
(**a**) and final (**b**) designs: with
the finite (*solid line*) and
infinite (*dashed line*) ground.
Specifications are shown with
the *thick solid line* (Ogurtsov
and Koziel 2011c)

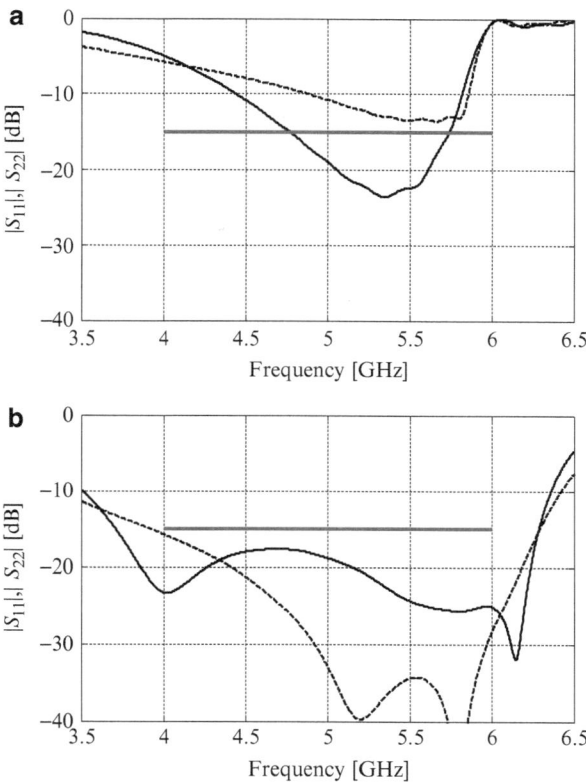

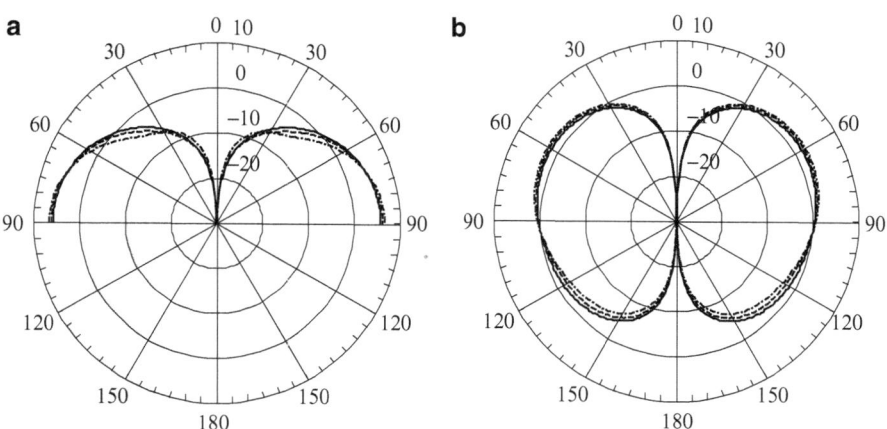

Fig. 7.16 Final design gain pattern [dBi]: (**a**) DRA with the infinite ground and (**b**) DRA with the
finite ground, at 4.5 GHz (*solid line*), 5.0 GHz (*dashed line*), and 5.5 GHz (*dash-dot lines*)
(Ogurtsov and Koziel 2011c)

One can infer from Fig. 7.15b that it would be possible to match the DRA better for each installation case separately. For example, the response of the DRA with the infinite ground could be shifted toward lower frequencies, which would result in better reflection response in the frequency band of interest. Our final design, however, is a compromise ensuring that the DRA satisfies the design requirements in the two installation scenarios.

7.4 Conclusions

In this chapter, we illustrated the application of two SBO approaches to DRA design optimization: SM combined with RSA (kriging) and AADS technique. Particularly space mapping can be considered as a general-purpose SBO approach that can be applied in cases where the discrepancy between the low- and high-fidelity models cannot be easily removed using simple means such as response correction or frequency scaling. The use of RSA is essential to reduce computational complexity of the design process, particularly when a large number of adjustable parameters are involved, which is the case for the DRA examples presented here.

It should be mentioned that other SBO techniques can also be applied for DRA optimization, including SPRP (Koziel and Ogurtsov 2012b), frequency scaling and output SM (Koziel and Ogurtsov 2012a), and a combination of SM and kriging with frequency scaling (Koziel and Ogurtsov 2011g).

Chapter 8
Surrogate-Based Optimization of Microstrip Broadband Antennas

In this chapter, we demonstrate the use of physics-based SBO techniques for the design of microstrip broadband antennas. A popular approach to make microstrip antenna responses broadband is to use thick multilayer substrates and extra, so-called parasitic, patches. For antennas of such a structure, the primary design tasks are matching in the frequency band of interest and maintaining the radiation response at required levels (e.g., peak gain, direction of maximal radiation, back-radiation level, radiation pattern, etc.)

Challenges common to all presented examples are the following:

- Simple analytical models of the microstrip broadband antennas are inaccurate—so they can be used only to set up initial designs.
- Accurate simulations are quite expensive and the simulated antenna response is the net effect of the radiator, parasitic patches, finite substrate/ground, and feed. Therefore, a realistic antenna model should include all these antenna parts and be simulated as a whole structure.
- The effect of a particular design variable is design dependent so that an improvement through a one-by-one parameter sweep is hardly feasible to achieve; therefore, automated tuning is the only reliable option to improve the design.

For details concerning background, formulation, and implementation of the SBO algorithms, an interested reader is encouraged to refer to Chapters 3 and 4 of this book.

8.1 Wideband Microstrip Antenna

Consider an antenna shown in Fig. 8.1 (Chen 2008, Koziel and Ogurtsov 2011c). Design parameters are $x = [l_1 \ l_2 \ l_3 \ l_4 \ w_2 \ w_3 \ d_1 \ s]^T$. A multilayer substrate is $l_s \times l_s$ ($l_s = 30$ mm). The antenna comprises (in the bottom-to-top order) metal ground, 0.813 mm-thick RO4003 layer; microstrip trace with width $w_1 = 1.1$ mm; 1.905 mm-thick RO3006 layer and a trace-to-driven patch via with radius $r_0 = 0.25$ mm; driven

S. Koziel and S. Ogurtsov, *Antenna Design by Simulation-Driven Optimization*, SpringerBriefs in Optimization, DOI 10.1007/978-3-319-04367-8_8, © Slawomir Koziel and Stanislav Ogurtsov 2014

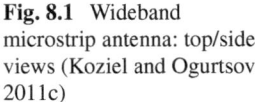

Fig. 8.1 Wideband microstrip antenna: top/side views (Koziel and Ogurtsov 2011c)

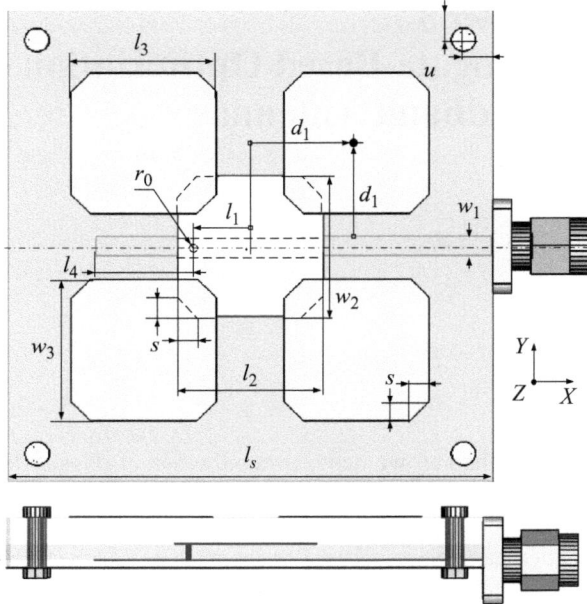

patch; 3. 048 mm-thick RO4003 layer; and four patches at the top. The antenna stack is fixed with four M1.6 bolts at the corners ($u = 3$ mm). Metallization is with thick 50 μm copper. Feeding is through an edge mount 50 Ω SMA connector with the 10 mm × 10 mm × 2 mm flange.

The design objective is $|S_{11}| \leq -10$ dB for 3.1 GHz to 4.8 GHz. Realized gain not less than 5 dB for the zero zenith angle is an optimization constrain over the frequency band. The initial design is $x^{init} = [-4\ 15\ 15\ 2\ 15\ 15\ 20\ 2]^T$ mm.

Both the high-fidelity model R_f (2,334,312 mesh cells at the initial design, 160 min of the evaluation time) and the low-fidelity model R_{cd} (122,713 mesh cells, 3 min of the evaluation time) are simulated using the CST MWS transient solver.

An algorithm exploiting coarse-discretization models should be designed to reduce not only the number of high-fidelity model evaluations but also low-fidelity ones. With this respect, space mapping may not be the best choice because of its parameter extraction step (Bandler et al. 2004a, b) so that it would be beneficial to skip the parameter extraction stage. One possibility of this is the SPRP technique which does not use any extractable parameters. Detailed formulation of the SPRP technique can be found in Sect. 4.3 of this book.

Here, the first step is to find a rough optimum of R_{cd}. With the use of the pattern search algorithm (Kolda et al. 2003), the approximate optimum is located at $x^{(0)} = [-4.91\ 15.15\ 15.07\ 2.56\ 14.21\ 14.23\ 21.07\ 2.67]^T$ mm. The computational cost of this step is 82 evaluations of R_{cd} which corresponds to about 1.5 evaluations of the high-fidelity model R_f. Figure 8.2(a) shows the responses of R_f at x^{init} and $x^{(0)}$, as well as the response of R_{cd} at $x^{(0)}$.

Fig. 8.2 Wideband microstrip antenna: (**a**) high-fidelity model response (*dashed line*) at the initial design x^{init}, and high- (*solid line*) and low-fidelity (*dotted line*) model responses at the approximate low-fidelity model optimum $x^{(0)}$; (**b**) high-fidelity model response at the final design (Koziel et al. 2012b)

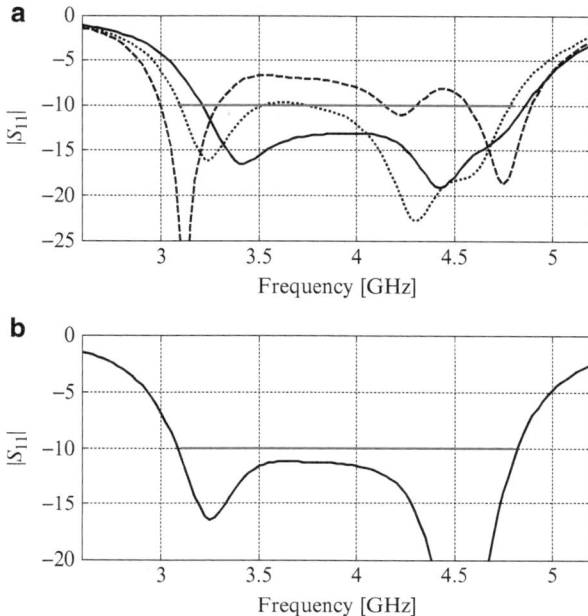

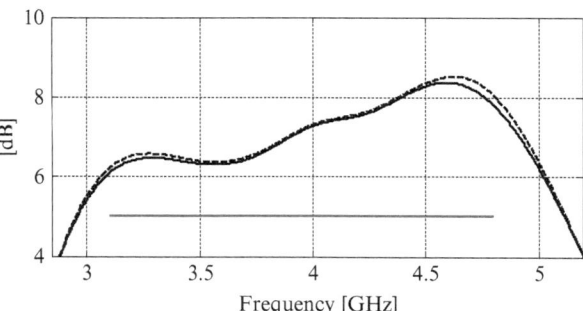

Fig. 8.3 Wideband microstrip antenna at the final design: realized gain for the zero zenith angle (*solid line*, XOZ co-pol.) and realized peak gain (*dashed line*). Design constrain is shown with the *horizontal line* at the 5 dB level (Koziel et al 2012b)

The final design $x^{(4)} = [-5.21 \; 15.38 \; 15.57 \; 2.58 \; 14.41 \; 13.73 \; 21.07 \; 2.067]^T$ mm with $|S_{11}| \leq -11$ dB for 3.1 GHz to 4.8 GHz shown in Fig. 8.2(b) is obtained after four iterations of the SPRP optimization. The antenna gain at the final design is shown in Fig. 8.3. The total design cost corresponds to about 10 evaluations of the high-fidelity model. Design cost summary is given with Table 8.1.

Another SBO approach which does not include the parameter extraction step is the variable-fidelity simulation-driven optimization (VFSDO) technique. VFSDO exploits a family of coarse-discretization models that are optimized sequentially, as well as the refinement step that uses an auxiliary response surface approximation model. An interested reader is referred to Sect. 4.7, where this technique is described in detail. The advantages of VFSDO include computational efficiency

Table 8.1 Wideband microstrip antenna: SPRP optimization cost (Koziel et al 2012b)

		Evaluation time	
Algorithm component	Model evaluations	Absolute (hours)	Relative to R_f
Evaluation of R_{cd}[a]	$289 \times R_{cd}$	14.4	5.4
Evaluation of R_f[b]	$5 \times R_f$	13.3	5.0
Total time	N/A	27.7	**10.4**

[a]Includes initial optimization of R_{cd} and optimization of SPRP surrogate
[b]Excludes evaluation of R_f at the initial design

Table 8.2 Wideband microstrip antenna: VFSDO cost (Koziel and Ogurtsov 2011c)

		Computational cost	
Algorithm component	Model evaluations	Absolute (hours)	Relative to R_f
Optimization of $R_{c.1}$	$125 \times R_{c.1}$	6.3	2.6
Optimization of $R_{c.2}$	$48 \times R_{c.2}$	14.4	5.4
Setup of model q	$17 \times R_{c.2}$	5.1	1.9
Evaluation of R_f[a]	$2 \times R_f$	5.3	2.0
Total time	N/A	31.1	**11.9**

[a]Excludes R_f evaluation at the initial design

(the high-fidelity model is only evaluated at the last stage of the optimization process), simplicity (no modifications of the models are necessary), and robustness.

The antenna of Fig. 8.1 was optimized using the VFSDO technique starting from the same initial design with the same design objective and constrain (Koziel and Ogurtsov 2011c). There are two coarse-discretization models: $R_{c.1}$ (122,713 mesh cells at the initial design) and $R_{c.2}$ (777,888 mesh cells at the initial design). The high-fidelity model R_f is also the same as with the SPRP technique. The evaluation times for $R_{c.1}$, $R_{c.2}$, and R_f are 3 min, 18 min, and 160 min at $x^{(\text{init})}$, respectively. All the models are evaluated using the time-domain solver of CST Microwave Studio. The final design $x^* = [14.87\ 13.95\ 15.4\ 13.13\ 20.87{-}5.90\ 2.88\ 0.68]^T$ mm is of similar quality, its $|S_{11}| \leq -11.5$ dB for 3.1 GHz to 4.8 GHz. The optimization cost (Table 8.2) corresponds to about 12 evaluations of the high-fidelity model R_f. Notice that although the total cost of VFSDO, with a particular realization of (Koziel and Ogurtsov 2011c), is higher, it is also 0.5 dB of $|S_{11}|$ better over the bandwidth of interest. Furthermore, computational costs with VFSDO can be substantially reduced with an algorithm of (Koziel and Ogurtsov 2011a, b, c, d) optimizing the low-fidelity models. For this particular antenna, twofold speedup has been demonstrated (Koziel and Ogurtsov 2011a, b, c, d).

8.2   Double-Ring Antenna

Consider a double-ring antenna (Kokotoff et al. 1999) shown in Fig. 8.4. It has three layers with permittivity of $\varepsilon_{r1} = 2.2$, $\varepsilon_{r2} = 1.07$, and $\varepsilon_{r3} = 2.2$ and loss tangent of 0.001 for all layers. The ground plane is modeled as infinite. All metal parts have

Fig. 8.4 Double-ring antenna: view

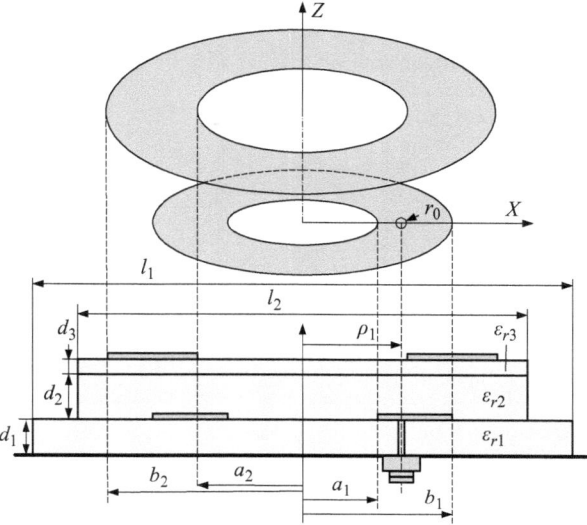

Fig. 8.5 Double-ring antenna: responses of the coarse-discretization model $R_{c.1}$ at the initial design $x^{(0)}$ (*dashed line*) and at the optimized design $x^{(1)}$ (*solid line*). Design specifications marked with the *horizontal line* (Koziel and Ogurtsov 2011c)

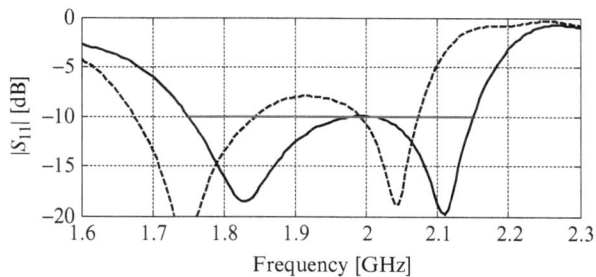

conductivity of copper, 5.8e7 S/m. Rings' thickness is 0.05 mm. Design variables are inner and outer radii of the rings, location of the feed's pin, thicknesses, and lateral extends of the first and second dielectrics: $x = [a_1 \ a_2 \ b_1 \ b_2 \ \rho_1 \ d_1 \ d_2 \ l_1 \ l_2]^T$ mm. The radius of the pin and thickness of the topmost dielectric are fixed to $r_0 = 0.325$ mm and $d_3 = 0.254$ mm.

Design specifications are $|S_{11}| \leq -10$ dB for 1.75–2.15 GHz. Requirement on the realized gain to be not less than 7 dB for the zero zenith angle and over the frequency band of interest is imposed as an optimization constrain. Design starts from $x^{init} = [10 \ 15 \ 30 \ 30 \ 20 \ 6 \ 8 \ 100 \ 100]^T$ mm.

To solve the problem, we adopt the VFSDO technique which uses two coarse-discretization models: $R_{c.1}$ and $R_{c.2}$. The evaluation times for $R_{c.1}$, $R_{c.2}$, and R_f are 2.5 min, 13 min, and 180 min at the initial design, respectively. All models are evaluated using the time-domain solver of CST Microwave Studio. The fine model is fed through full-wave 50 Ω coaxial port, whereas the coarse-discretization models are excited by 50 Ω discrete source at the gap (0.5 mm) between the ground and the pin.

Figure 8.5 shows the responses of $R_{c.1}$ at $x^{(0)}$ and at its optimal design $x^{(1)}$. Figure 8.6 shows the responses of $R_{c.2}$ at $x^{(1)}$ and at its optimized design $x^{(2)}$.

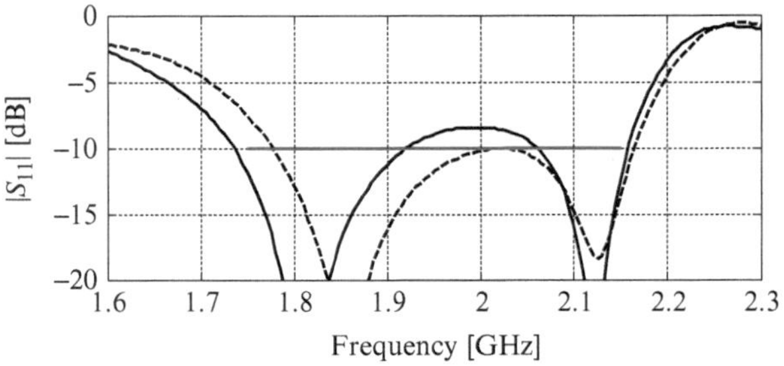

Fig. 8.6 Double-ring antenna: responses of the coarse-discretization model $R_{c.2}$ at $x^{(1)}$ (*dashed line*) and at its optimized design $x^{(2)}$ (*solid line*). Design specifications marked with the *horizontal line* (Koziel and Ogurtsov 2011c)

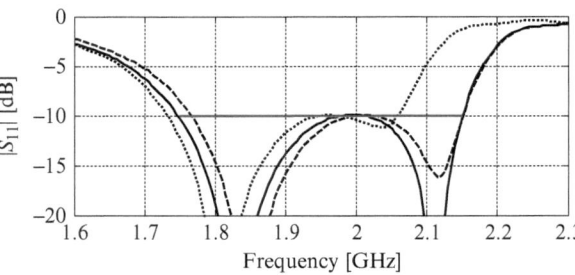

Fig. 8.7 Double-ring antenna: responses of the original, high-fidelity model R_f at $x^{(0)}$ (*dotted line*), at $x^{(2)}$ (*dashed line*), and at the refined final design x^* (*solid line*). Design specifications marked with the *horizontal line* (Koziel and Ogurtsov 2011c)

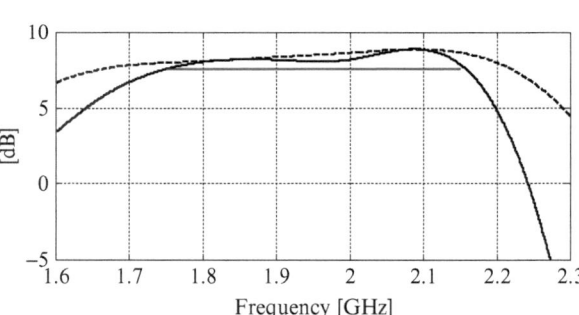

Fig. 8.8 Double-ring antenna at the final design, far-field response for the zero zenith angle: IEEE gain (*dashed line*); realized gain (*solid line*). Design constrain imposed on the realized gain is marked with the *horizontal line* (Koziel and Ogurtsov 2011c)

Table 8.3 Double-ring antenna: VFSDO cost summary (Koziel and Ogurtsov 2011c)

Algorithm component	Model evaluations	Computational cost	
		Absolute (hours)	Relative to R_f
Optimization of $R_{c.1}$	$178 \times R_{c.1}$	7.4	2.5
Optimization of $R_{c.2}$	$83 \times R_{c.2}$	18.0	6.0
Setup of model q	$19 \times R_{c.2}$	4.1	1.3
Evaluation of R_f^a	$3 \times R_f$	9.0	3.0
Total time	N/A	38.5	**12.8**

[a]Excludes R_f evaluation at the initial design

Figure 8.7 shows the responses of R_f at $x^{(0)}$, at $x^{(2)}$, and at the refined design $x^* = [10.09\ 11.75\ 28.81\ 30.32\ 20.93\ 6.85\ 8.50\ 104.50\ 103.75]^T$ mm ($|S_{11}| \le -10$ dB for 1.75 GHz to 2.15 GHz) obtained in three iterations of the refinement step of VFSDO; cf. (4.24)–(4.26). Gain responses of the final design are shown in Fig. 8.8. The optimization cost is shown in Table 8.3.

8.3 Microstrip Antenna with U-Shape Parasitic Patches

As the last example, consider a coax-fed microstrip antenna shown in Fig 8.9. The antenna is on 3.81 mm-thick Rogers TMM4 ($\varepsilon_1 = 4.5$ at 10 GHz). The TMM4 lateral dimensions are $l_x = l_y = 6.75$ mm. The ground plane is of infinite extends. The feed probe diameter is 0.8 mm. The connector's inner conductor is 1.27 mm in diameter.

Design specifications are $|S_{11}| \leq -10$ dB for 5 GHz to 6 GHz. Design variables are $x = [a\ b\ c\ d\ e\ l_0\ a_0\ b_0]^T$. The initial design is $x^{(0)} = [6\ 12\ 15\ 1\ 1\ 1\ 1\ 1\ -4]^T$ mm.

We use this example to emphasize the importance of appropriate selection of the low-fidelity model. In order to do that, the antenna is optimized three times, using the same SBO algorithm working with three different coarse-discretization models: R_{c1} (41,496, 1 min), R_{c2} (96,096, 3 min), and R_{c3} (180,480, 6 min). We investigate the performance of the SBO algorithm working with these models in terms of the computational cost and the quality of the final design. The high-fidelity model R_f (704,165 mesh cells, evaluation time 60 min) and the low-fidelity models are evaluated with CST MWS transient solver.

Figure 8.10 shows the responses of all the models at the approximate optimum of R_{c1}. The major misalignment between the responses is due to the frequency shift

a   b

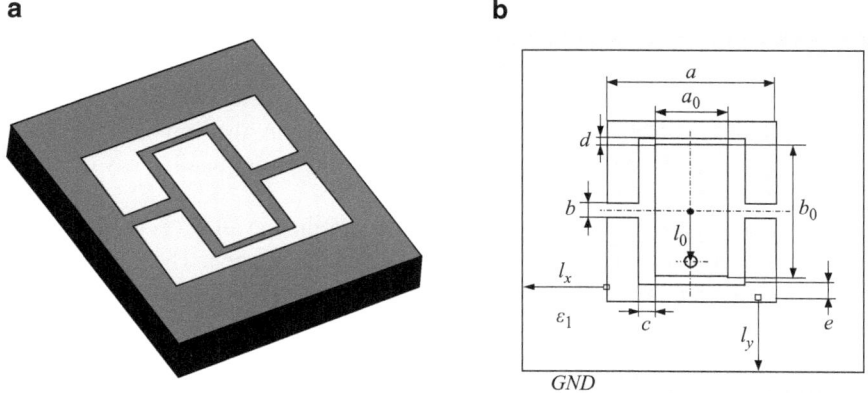

Fig. 8.9 Microstrip antenna: (**a**) 3D view and (**b**) layout top view (Koziel and Ogurtsov 2012b)

Fig. 8.10 Antenna of Fig. 8.9: model responses at the approximate optimum of R_{c1}: R_{c1} (*dotted line*), R_{c2} (*dash-dot lines*), R_{c3} (*dashed line*), and R_f (*solid line*) (Koziel and Ogurtsov 2012b)

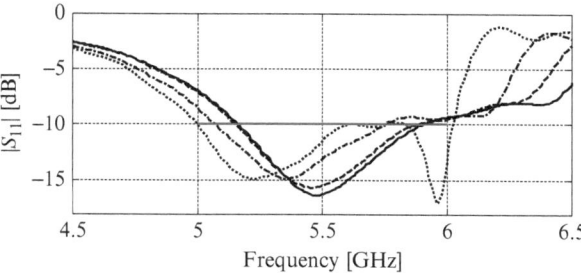

Table 8.4 Microstrip antenna: design results and costs (Koziel and Ogurtsov 2012b)

Low-fidelity model	Design cost: the number of model evaluations[a]		Relative design cost[b]	Max$|S_{11}|$ for 2–8 GHz at final design
	R_c	R_f		
R_{c1}	385	6	12.4	−8.0 dB
R_{c2}	185	3	12.3	−10.0 dB
R_{c3}	121	2	14.1	−10.7 dB

[a]Number of R_f evaluations is equal to the number of the SBO algorithm iterations
[b]Equivalent number of R_f evaluations includes evaluation at the initial design

Fig. 8.11 Antenna of Fig. 8.9: the high-fidelity model R_f response at the final design found using the low-fidelity model R_{c3} (Koziel and Ogurtsov 2012b)

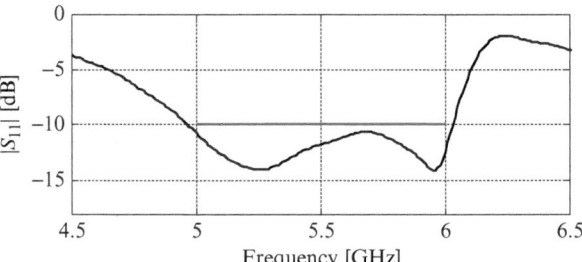

so that the surrogate is created here using the frequency scaling technique as well as additive output space mapping (Bandler et al. 2004a, b). Frequency scaling is implemented as described in Sect. 3.2.2 (cf. (3.14), (3.15)). It is realized using appropriate interpolation/extrapolation schemes in order to obtain the low-fidelity model response at any frequency necessary. As a result, it does not involve any extra EM simulations.

The results, Table 8.4 and Fig. 8.10, indicate that the model R_{c1} is too inaccurate and the SBO process using it fails to find a satisfactory design. The designs found with models R_{c2} and R_{c3} satisfy the specifications and the cost of the SBO process using R_{c2} is slightly lower than that with R_{c3}.

The problem of selecting low-fidelity models for a particular SBO algorithm to ensure its efficiency both in terms of CPU costs and quality of the final design is addressed in more detail in Chapter 13 of this book (Fig. 8.11).

8.4 Conclusions

Similarly as for UWB antennas, surrogate-based optimization of broadband antennas can be performed using various methods. In case the discrepancy between the low- and high-fidelity models can be clearly identified (e.g., as mostly frequency misalignment or vertical shift), simple means such as frequency scaling or additive response correction can be utilized (cf. Sect. 8.3). If the antenna response is characterized by certain distinctive features (e.g., local reflection minima, etc.), methods exploiting this information such as SPRP are very convenient (cf. Sect. 8.1). The VFSDO algorithm demonstrated for both wideband microstrip antenna of Sect. 8.1 and double-ring antenna of Sect. 8.2 while not being the fastest, is usually quite robust and recommended in case of doubts regarding the method selection or if the most suitable type of low-fidelity model correction cannot be easily inferred from visual inspection of the antenna responses.

Chapter 9
Simulation-Driven Antenna Array Optimization

In this chapter we apply the SBO methodology to simulation-driven design of planar antenna arrays. Reliable designs of planar arrays are challenging due to the time-consuming high-fidelity electromagnetic (EM) simulations necessary to evaluate both radiation and reflection responses of the realistic array model (Volakis 2007). In addition, antenna array designs involve large numbers of design variables, including dimensions of elements, location of feeds, spacings, excitation amplitudes and/or phases, finite dimensions of substrates, and grounds. Models based on the single element radiation response combined with the analytical array factor (Balanis 2005) do not produce accurate radiation responses in the directions off the main beam and fail to account for inter-element coupling. Therefore, the use of full-wave EM models for the entire array is necessary. Such models, however, are computationally expensive when accurate, and conducting array design through simulation-driven optimization might be prohibitively expensive in terms of the CPU time. To alleviate this difficulty and speed up the design optimization process, we exploit the SBO approach.

Array design normally comprises two major steps: adjusting of the radiation response, e.g., directivity pattern, and adjusting the reflection response. The use of surrogate models can be beneficial at both of these two steps.

In Sect. 9.1 we demonstrate the design of a 5×5 array of microstrip antennas using a low-fidelity coarse-discretization model of the entire array exploited in through the design process. In Sect. 9.2 we demonstrate design of a 7×7 array of microstrip antennas using two surrogate models, one based on the single element radiation response combined with the analytical array factor and the other based on the coarse-discretization model of the entire array.

9.1 5×5 Antenna Array

Consider a rectangular planar array (Fig. 9.1) comprising 25 identical microstrip patches residing on a 1.58 mm-thick RT/duroid 5880 is the substrate. Each patch is feed by a probe in the 50Ω environment. Initial dimensions of elements are 11 mm by 9 mm.

S. Koziel and S. Ogurtsov, *Antenna Design by Simulation-Driven Optimization*,
SpringerBriefs in Optimization, DOI 10.1007/978-3-319-04367-8_9,
© Slawomir Koziel and Stanislav Ogurtsov 2014

Fig. 9.1 Microstrip antenna array. The symmetry plane (magnetic wall) is shown with the *vertical dash line* (Koziel and Ogurtsov 2013d)

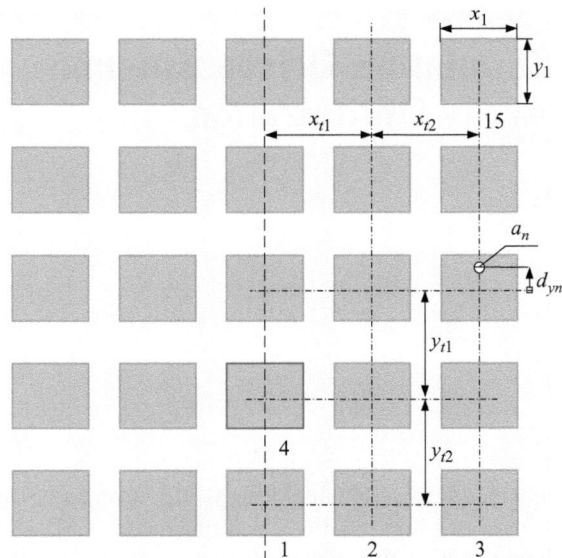

Lateral extensions of the substrate and the ground are set to a half of the patch size in a particular direction. The symmetry is imposed as shown in Fig. 9.1. Consider also the following design requirements: the array is to operate at 10 GHz, be of linear polarization, and have the direction of the maximum radiation perpendicular to the plane of the array; peak directivity should be about 20 dBi; the minor lobes are to be below −20 dB; and returning signals at the element feeds should be lower than −10 dB.

The main design challenge is a large number of variables. Even with the imposed symmetry and restriction to adjust distances between patches (x_{t1}, x_{t2}, y_{t1}, y_{t2}), patch size (x_1, y_1), probe offsets (d_{y1}, d_{y2}, ... d_{y15}), and incident excitation amplitudes (a_1,...a_{15}), the number of variables is still 36. With the simulation time of the high-fidelity model $\boldsymbol{R}_f$ being around 20 min with CST MWS software (CST MWS 2013) and on a 2.53 GHz quad-core Intel Xeon processor with 6 Gb RAM, direct optimization turns to be hardly feasible. To alleviate this difficulty, we exploit the SBO methodology with an auxiliary low-fidelity model $\boldsymbol{R}_c$, which is also evaluated in CST MWS but with a coarser mesh (evaluation time around 1 min). This model $\boldsymbol{R}_c$ describes the directivity pattern quite accurately within the main beam, but it is not particularly good in representing the reflection response. It is worth to note that although we use a term of reflection response and notation $|S_k|$ referring returning signals at the feed points, these signals include the effect of coupling due to simultaneous excitation of the elements, i.e., $|S_k|$ refer to so-called active S-parameters (CST MWS 2013).

We split the design variable vector $\boldsymbol{x}$ into two parts: $\boldsymbol{x} = [\boldsymbol{x}_p^{\mathrm{T}}\ \boldsymbol{x}_m^{\mathrm{T}}]^{\mathrm{T}}$, where $\boldsymbol{x}_p = [x_{t1}$ $x_{t2}\ y_{t1}\ y_{t2}\ x_1\ y_1\ a_1...a_{15}]$ comprises variables used to optimize the array directivity pattern and $\boldsymbol{x}_m = [d_{y1}\ d_{y2} ... d_{y15}]$ comprises variables used to adjust the reflection response.

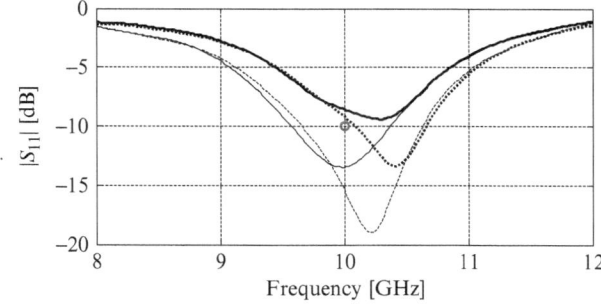

Fig. 9.2 Reflection responses of R_f (selected port) (*thick solid line*) and R_c (*thin solid line*) at $x = [(x_p^*)^T (x_{m.0})^T]^T$ and at a design with a d_{yk}—variable corresponding to port k perturbed by certain Δd_{yk} (*thick* and *think dotted lines*). Based on these R_c responses and R_f at $[(x_p^*)^T (x_{m.0})^T]^T$, a proper perturbation for d_{yk} is found as described in Step 2. Note that additional "horizontal" correction of this response may be necessary which is realized as described in Step 3. A *circle* denotes design specifications for reflection coefficient (Koziel and Ogurtsov 2013d)

Having this in mind, the following 3-step design procedure has been developed (Koziel and Ogurtsov 2013d):

Step 1: Optimize the directivity pattern of the low-fidelity model R_c using x_p with fixed $x_m = x_{m.0}$ (the initial value); the optimized x_p will be referred to as x_p^*. Optimization of R_c in this stage is realized using auxiliary 1st-order response surface models constructed using large-step design perturbations and trust-region framework to ensure convergence.

Step 2: Evaluate model R_f at $x = [(x_p^*)^T (x_{m.0})^T]^T$. Use R_c to estimate the necessary changes in x_m to improve reflection responses. Here, it is assumed that a small change (tuning) of a given x_m component noticeably affects the reflection of the corresponding patches and not those of the others. It has been verified with numerical experiments that this assumption is satisfied for the structure under design for the used range of design variables. The methodology we use is the following: (i) evaluate model R_c at $x = [(x_p^*)^T (x_{m.0})^T]^T$ and at the two perturbed designs varied by $\pm\Delta d_y$ corresponding to a reflection response that does not satisfy matching requirements (cf. Fig. 9.2) and (ii) using interpolation of the data obtained in (i), estimate the change of d_y that gives reasonable change of the response (this takes into account the fact that responses of R_f and R_c are shifted both in frequency and amplitude). The modified vector x_m will be referred to as x_m^*.

Step 3: Evaluate R_f at $x = [(x_p^*)^T (x_m^*)^T]^T$; adjust the global parameter y_1 (patch length) to shift the matching responses in frequency as necessary. The change of y_1 is estimated using evaluation of R_c at $x = [(x_p^*)^T (x_m^*)^T]^T$ and the two perturbed designs obtained by changing y_1 and interpolating the results. The design, obtained after this step, will be referred to as x^*.

It should be noted that the low-fidelity model is used as much as possible, and the high-fidelity model is only evaluated in Step 2 (once) and in Step 2 (twice).

Fig. 9.3 Radiation response of high-fidelity model $\boldsymbol{R}_f$ at the initial design $\boldsymbol{x}^{(0)}$ (uniform array): directivity pattern cuts in the E and H planes at 10 GHz (Koziel and Ogurtsov 2013d)

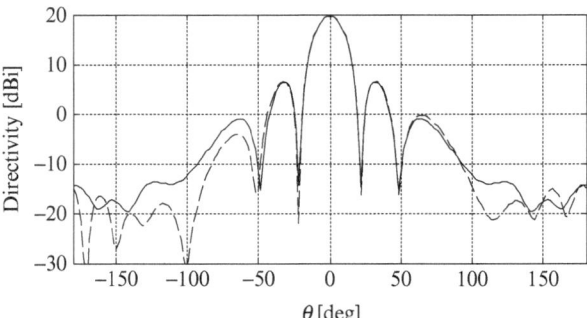

Fig. 9.4 Reflection of high-fidelity model $\boldsymbol{R}_f$ at the initial design with solid lines and $|S_{11}|$ of the single isolated element with *thick dash line* (Koziel and Ogurtsov 2013d)

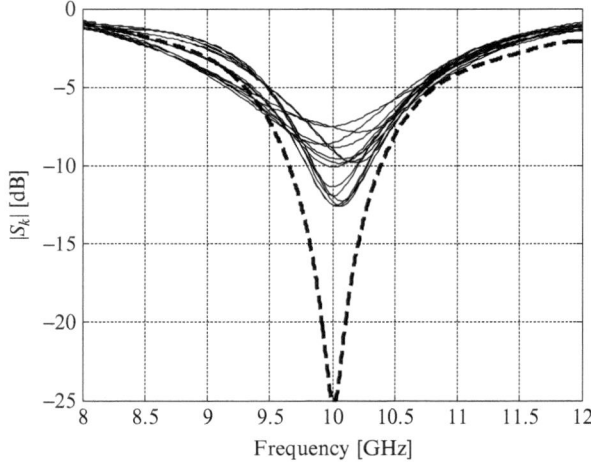

Consider responses of the array at the initial uniform design $\boldsymbol{x}^{(0)} = [x_{t1}\ x_{t2}\ y_{t1}\ y_{t2}\ x_1\ y_1\ a_1...a_{15}\ d_{y1}\ d_{y2}\ ...\ d_{y15}] = [16\ 16\ 16\ 16\ 11\ 9\ 1\ ...\ 1\ 2.9\ ...\ 2.9]^T$ where dimensions are in mm and incident excitation amplitudes are unitless. The responses are shown in Figs. 9.3 and 9.4.

Design specifications for Step 1 (directivity pattern optimization) are the following: minimize directivity (in the minimax sense) off the main beam of design $\boldsymbol{x}^{(0)}$, i.e., for the zenith angles off the sector $[-21.5°, 21.5°]$. Step 1 (optimization of the coarse model for pattern) results in design $\boldsymbol{x}_p^* = [16.363\ 16.588\ 16.498\ 16.910\ 11.072\ 8.926\ 0.9845\ 0.4529\ 0.3718\ 0.9873\ 0.9748\ 0.4500\ 0.9970\ 0.9754\ 0.9919\ 0.9548\ 0.9369\ 0.5503\ 0.9999\ 0.4671\ 0.3621]^T$. Responses of the array after step 1 are shown in Figs. 9.5 and 9.6. The cost of Step 1 is $182 \times \boldsymbol{R}_c$.

At Step 2 (matching correction I), we change d_{yk} for ports where matching is not sufficient (i.e., > -10 dB). For ports 4, 7, 8, and 10 (see Fig. 9.5), the feed location is increased to 3.4 mm. The cost of step 2 is $8 \times \boldsymbol{R}_c + 1 \times \boldsymbol{R}_f$.

At Step 3 (matching correction II), one changes the global parameter y_1 to 9.1 mm to move reflection responses to the left in frequency. This step costs $2 \times \boldsymbol{R}_c + 2 \times \boldsymbol{R}_f$.

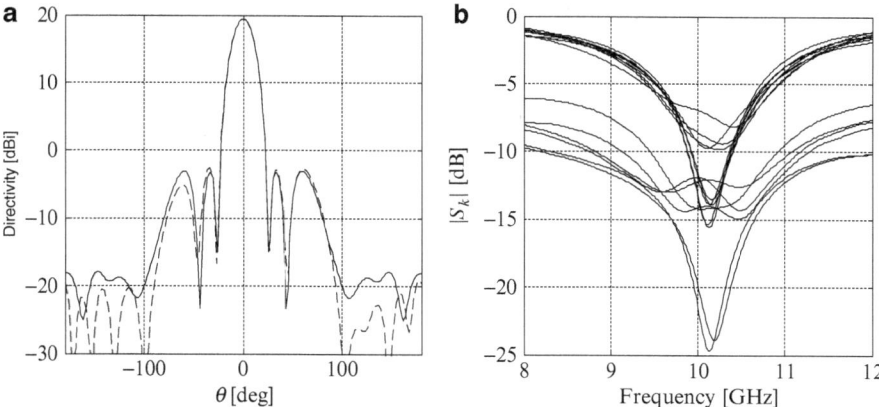

Fig. 9.5 High-fidelity model R_f after Step 3 (directivity pattern optimization): (**a**) radiation response, (**b**) reflection response. Steps 2 and 3 are yet to be done (Koziel and Ogurtsov 2013d)

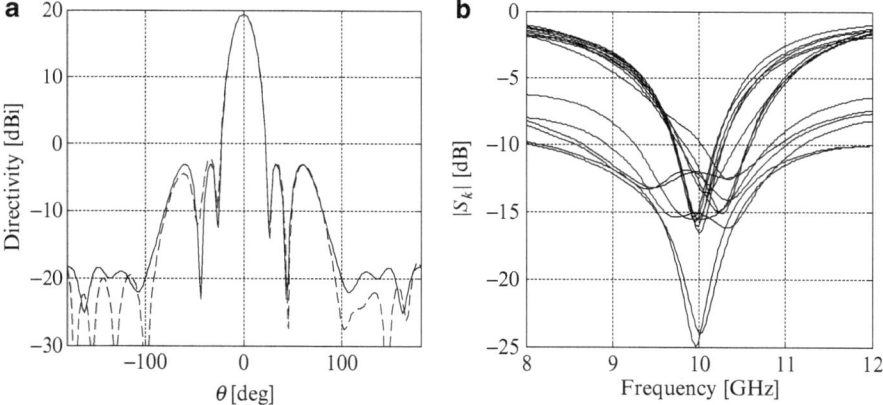

Fig. 9.6 High-fidelity model R_f after Step 3, i.e., at the final design: radiation response in the E and H plane at 10 GHz; (**b**) refection (Koziel and Ogurtsov 2013d)

The total cost of the design optimization process is $192 \times R_c + 3 \times R_f = 12.5 \times R_f$, i.e., it is equivalent in time to only 12.5 high-fidelity simulations of the entire structure. Responses of the final design are shown in Fig. 9.6.

9.2 Optimization of a 7×7 Array Using Analytical and Discrete Models

As it is seen from the cost budget of the previous example, a significant part (i.e., $8.5 \times R_f$), of the total design cost (i.e., $12.5 \times R_f$), is associated with Step 1 (directivity pattern adjustment), namely, it comes from multiple evaluation of the coarse-discretization model of the entire array.

In order to reduce the cost of the radiation response optimization step, we utilize an analytical model of the array R_a, directivity $D_a(\theta, \phi) \sim D_e(\theta, \phi) \cdot A(\theta, \phi)$, which embeds the EM-simulated radiation response of the single microstrip patch antenna $D_e(\theta, \phi)$ and analytical array factor $A(\theta, \phi)$ (Balanis 2005). The coarse-discretization model R_{cd} of the entire array is also used as an auxiliary tool to adjust the reflection response. The design process exploits the surrogate-based optimization (SBO) approach, where direct optimization of the array is replaced by iterative correction and adjustment of the auxiliary models R_a and R_{cd}. The design procedure consists of the following two stages (Koziel and Ogurtsov 2013e):

Stage 1 (pattern optimization): In this stage, the design variables x are optimized in order to reduce the side low level according to the specifications of Sect. 9.2. Starting from the initial design $x^{(0)}$, the first approximation $x^{(1)}$ of the optimum design is obtained by optimizing the analytical model R_a. Further approximations $x^{(i)}$, $i = 2, 3, \ldots$, are obtained as

$$x^{(i)} = \arg\min\left\{ x : R_a(x) + \left[R_f\left(x^{(i-1)}\right) - R_a\left(x^{(i-1)}\right) \right] \right\}, \tag{9.1}$$

i.e., by optimizing the analytical model R_a corrected using output space mapping (Koziel et al 2008b) so that it matches the high-fidelity model exactly at the previous design $x^{(i-1)}$. In practice, only two iterations are usually necessary to yield a satisfactory design. Note that each iteration of the above procedure requires only one evaluation of the high-fidelity model R_f. Response correction of the analytical model R_a is necessary because—as opposed to the coarse-discretization model R_{cd} used in the previous example to optimize the array pattern—the model R_a does not represent the pattern sufficiently well for radiation directions off the main beam.

Stage 2 (reflection adjustment): In this stage, the coarse-discretization model R_{cd} is used to correct the reflection of the array. Again, although we use the term "reflection response" and $|S_k|$ referring to returning signals at the feed points (ports), these signals include the effect of coupling due to simultaneous excitation of the elements. The analytical model R_a does not give any information about the array reflection, so that only the pattern can be considered in Stage 1. The array reflection has to be corrected in order to satisfy the requirement of -10 dB levels for returning signals at all ports.

In practice, after optimizing the pattern, the reflection responses are slightly shifted in frequency so that the minima of $|S_k|$ are not exactly at the required frequency (here, 10 GHz). The reflection responses can be shifted in frequency by adjusting the size of the patches, y_1 here. In order to find the appropriate change of y_1, we use the coarse-discretization model R_{cd}. Because both R_f and R_{cd} are evaluated using the same EM solver, we assume that the frequency shift of reflection responses is similar for both models under the same change of the variable y_1, even though responses themselves are not identical for R_f and R_{cd} (in particular, they are shifted in frequency and the minimum levels of $|S_k|$ are typically different). By performing perturbation of y_1 using R_{cd}, one can estimate the change of y_1 in R_f necessary to obtain the required frequency shift of its reflection responses. This change would

Fig. 9.7 Array of 49 microstrip patches: front view. Symmetry (magnetic) plane is shown with the *vertical dash line* at the center (Koziel and Ogurtsov 2013e)

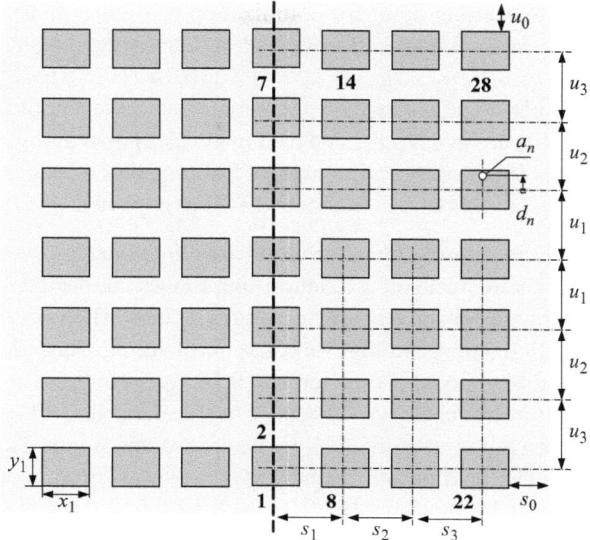

normally be very small so that it would not affect the array pattern in a substantial way. The computational cost of reflection adjustment using the method described here is only one evaluation of the high-fidelity model and one evaluation of the coarse-discretization model $\boldsymbol{R}_{cd}$.

In case of severe mismatch, the feed offsets d_n can also be used to adjust reflection, however, it was not necessary in the design cases considered here.

The design procedure described above is applied to an array of Fig. 9.7. The array is to operate at 10 GHz with a linear polarization in the E-plane. Each patch is fed by a probe in the 50 Ω environment. Initial dimensions of elements, microstrip patches, are 11 mm by 9 mm; a grounded layer of 1.58 mm-thick RT/duroid 5880 is the substrate; the extension of the substrate/ground s_0 and u_0 is set to 15 mm.

The design tasks are to have (a) the lobe level below −20 dB for zenith angles off the main beam with the null-to-null width of 32°, i.e., off the sector of [−16°, 16°]; (b) the peak directivity about 20 dBi; (c) the direction of the maximum radiation perpendicular to the plane of the array; and (d) returning signals lower than −10 dB, all at 10 GHz.

The simulation time of the high-fidelity model of the array, $\boldsymbol{R}_f$, is around 30 min using the CST MWS transient solver. Even though we impose symmetry and, therefore, restrict ourselves to adjusting spacing (s_1, s_2, s_3, u_1, u_2, u_3), patch size (x_1, y_1), location of fed probes ($d_1, \ldots d_{28}$), amplitudes ($a_1, \ldots a_{28}$), and phases ($b_1, \ldots b_{28}$) of the incident signals, the number of variables is still large for direct optimization. Therefore, we consider two design optimization cases: a design with nonuniform amplitude (and uniform phase) excitation with the design variables being $\boldsymbol{x} = [s_1 \ s_2 \ s_3 \ u_1 \ u_2 \ u_3 \ x_1 \ y_1 \ a_1 \ldots a_{28} \ d_1 \ldots d_{28}]^T$ and with nonuniform phase (uniform amplitude) excitation with $\boldsymbol{x} = [s_1 \ s_2 \ s_3 \ u_1 \ u_2 \ u_3 \ x_1 \ y_1 \ b_1 \ldots b_{28} \ d_1 \ldots d_{28}]^T$. The coarse-discretization model $\boldsymbol{R}_{cd}$ of the entire array runs in about 1 min.

A starting point for optimization is a uniform array design. The array spacings are easily found using model R_a assuming them equal to each other. $x^{(0)} = [s_1 \, s_2 \, s_3 \, u_1$ $u_2 \, u_3 \, x_1 \, y_1 \, a_1 \, ... \, a_{28} \, d_1 \, ... \, d_{28}]^T = [16 \, 16 \, 16 \, 16 \, 16 \, 11 \, 9 \, 1 \, ... \, 1 \, 2.9...2.9]$ where all dimensional parameters are in mm, excitation amplitudes are normalized, and phase shifts are in degrees. The feed offset d_n, shown in Fig. 9.7, is 2.9 mm for all patches; it is obtained by optimizing the EM model of the single patch antenna. The side lobe level of this design is about −13 dB as expected, and the peak directivity is 22.7 dBi.

Optimization with nonuniform amplitude excitation. The design has been carried out with incident excitation amplitudes as design variables. Maximum allowed array spacings were restricted to 20 mm. The cost of Stage 1, directivity pattern optimization, is only 3 evaluation of R_f. At Stage 2, we change the y-size of the patches, global parameter y_1 to 9.14 mm in order to move reflection responses to the left in frequency y_1. Offsets d_n of the elements still violating the specification have been adjusted individually. The cost of this step is $5 \times R_{cd} + 1 \times R_f$.

The final design is found at $x^* = [s_1 \, s_2 \, s_3 \, u_1 \, u_2 \, u_3 \, x_1 \, y_1 \, a_1 \, ... \, a_{28}]^T = [15.97 \, 17.35$ $20.00 \, 14.38 \, 17.98 \, 19.99 \, 11.00 \, 9.14 \, 0.922 \, 0.787 \, 1.000 \, 0.835 \, 0.953 \, 0.779 \, 0.770$ $0.958 \, 0.966 \, 1.000 \, 0.810 \, 0.963 \, 0.989 \, 0.925 \, 0.452 \, 0.620 \, 0.832 \, 0.842 \, 0.814 \, 0.631$ $0.576 \, 0.072 \, 0.752 \, 0.697 \, 0.872 \, 0.821 \, 0.703 \, 0.037]^T$. Most probe offsets d_n have been left of the initial design value, 2.9 mm, except four adjusted to $d_4 = d_{11} = d_{18} = 3.9$ mm and $d_{10} = 3.4$ mm. The radiation response and reflection response of the final design are shown in Fig. 9.8. The side lobe level of this design x^* is under −20 dB and the peak directivity of x^* is 22.9 dBi. The total cost of optimization is only about $5 \times R_f$.

Optimization with nonuniform phase excitation. Another case has been considered with the excitation phase shifts as design variables and spacings restricted to 20 mm. The final design is at $x^* = [s_1 \, s_2 \, s_3 \, u_1 \, u_2 \, u_3 \, x_1 \, y_1 \, b_1 \, ... \, b_{28}]^T = [15.00 \, 15.00 \, 20.00 \, 15.15$ $5.46 \, 19.95 \, 11.00 \, 9.10 \, 0 \, 8.6 \, -6.3 \, 1.1 \, 4.3 \, 2.6 \, 3.1 \, 33.3 \, 0.3 \, 11.0 \, -4.9 \, 5.3 \, -14.6 \, 45.7$ $-60.7 \, 17.4 \, 5.8 \, 29.6 \, -7.0 \, 39.4 \, -48.9 \, -17.7 \, 46.5 \, -13.8 \, 22.5 \, -1.65 \, 47.9 \, -38.9]^T$ where the phase shifts are in degrees and given relatively the first element. Its responses are shown in Fig. 9.9. The side lobe level of this design x^* is under −17 dB; the peak directivity of $x^{(0)}$ is 22.2 dBi; return signals $|S_k|$ are higher than in the previous case; and their suppression should be addressed with design of the feed network. The total design cost is similar as for the previous example, about $5 \times R_f$.

9.3   Discussion and Conclusion

The considerations and results presented in this chapter illustrate the use of surrogate-based optimization for the design of antenna arrays. While the specific approaches adopted here are not quite standard (in the SBO sense and the optimization algorithms as presented, e.g., in Chap. 4), they demonstrate that using lower-fidelity models combined with simple correction techniques, problem decomposition, as well as "heuristic" approaches (particularly, using the sensitivity analysis of the low-fidelity model to estimate the necessary adjustments of the high-fidelity one), may lead to substantial reduction of the design cost. For both considered arrays, the

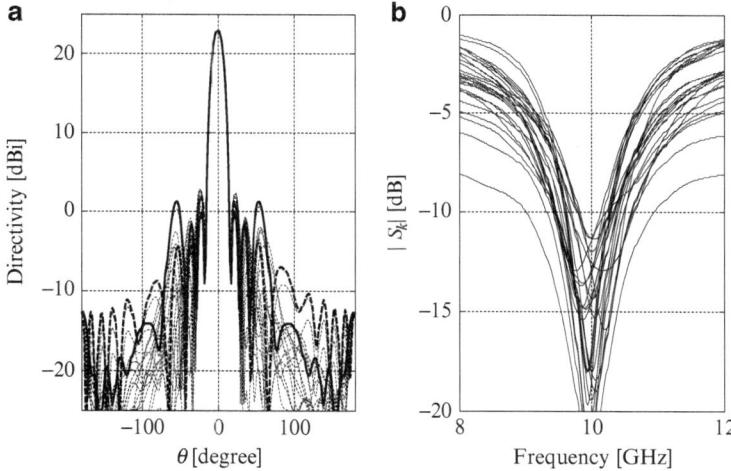

Fig. 9.8 Array optimized with nonuniform amplitude excitation and spacings constrained by 20 mm: (**a**) directivity pattern cuts for $\phi = 0°$, $5°$,..., $90°$ where H-plane cut ($\phi = 0°$) with (*solid line*) and E-plane cut ($\phi = 90°$) with (*dashed line*); (**b**) reflection (Koziel and Ogurtsov 2013e)

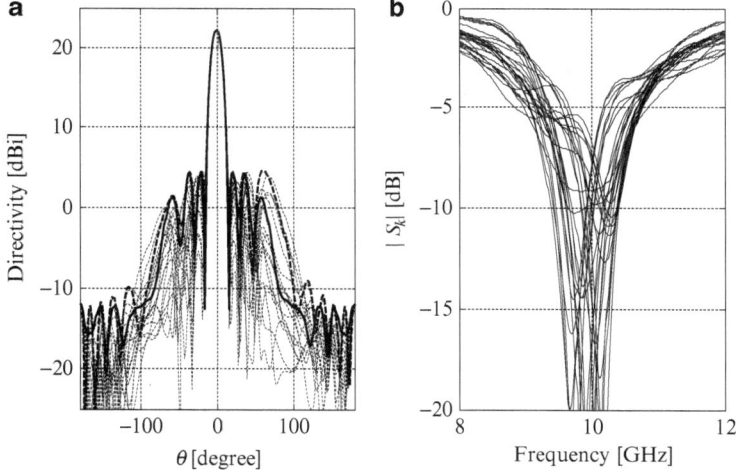

Fig. 9.9 Array optimized with nonuniform phase excitation and spacings constrained by 20 mm: (**a**) directivity pattern cuts for $\phi = 0°$, $5°$,..., $90°$ where H-plane cut ($\phi = 0°$) with (*solid line*) and E-plane cut ($\phi = 90°$) with (*dashed line*); (**b**) reflection (Koziel and Ogurtsov 2013e)

satisfactory designs were obtained at the cost corresponding to a few evaluations of the high-fidelity model, despite the fact that the numbers of designable parameters were quite large. At the same time, the design procedures discussed here indicate that successful utilization of the SBO paradigm requires certain insight into the problem at hand as well as experience in numerical modeling and optimization. This is one of the reasons why SBO, despite its huge potential, has not yet been widely adopted by engineers and designers.

Chapter 10
Antenna Optimization with Surrogates and Adjoint Sensitivities

Improving robustness and computational efficiency of simulation-driven design is possible with adjoint sensitivity that allows obtaining derivative information of the system of interest with little or no extra computational cost (Nair and Webb 2003; El Sabbagh et al. 2006; Kiziltas et al. 2003; Uchida et al. 2009; Bakr et al. 2011). Until recently, adjoint sensitivities were not implemented in commercial software packages which means that they were not accessible for most engineers and designers. The situation changed a few years ago when this technology has become available for instance in CST Microwave Studio (CST, 2013).

Using sensitivity information allows substantial enhancement of surrogate-based optimization schemes. First of all, the surrogate model itself can be constructed to be first-order consistent with the high-fidelity model (Alexandrov and Lewis 2001) so that the SBO algorithm becomes globally convergent in a classical sense when embedded in the trust-region framework (Conn et al. 2000). Cheap sensitivity through adjoints may also be used to speed up the surrogate model optimization step as well as the surrogate model parameter extraction process (if any), thus reducing the overall design time.

In this chapter, we review several techniques that exploit adjoint sensitivity in order to speed up the simulation-driven antenna design. These techniques include gradient-based search methods embedded in trust-region framework, as well as surrogate-based methods, specifically space mapping (Koziel et al. 2008a) and manifold mapping (Echeverria and Hemker 2005), enhanced by adjoint sensitivity in order to improve their convergence properties and reduce the computational cost of surrogate model optimization step. The efficiency of the presented approaches is demonstrated using several designs. A performance comparison with other optimization techniques, including Matlab's *fminimax* (Matlab 2012), is also provided.

S. Koziel and S. Ogurtsov, *Antenna Design by Simulation-Driven Optimization*,   93
SpringerBriefs in Optimization, DOI 10.1007/978-3-319-04367-8_10,
© Slawomir Koziel and Stanislav Ogurtsov 2014

10.1 Surrogate-Based Optimization with Adjoint Sensitivity

In this section, we recall the generic surrogate-based optimization scheme and discuss the convergence safeguard using trust-region framework and surrogate and high-fidelity model consistency. Various ways of ensuring consistency conditions using sensitivity are discussed in Sects. 10.2–10.4.

10.1.1 Generic Surrogate-Based Optimization Algorithm

A generic surrogate-based optimization algorithm generates a sequence of approximate solutions to (1), $x^{(i)}$, as follows (Koziel and Yang 2011; Forrester and Keane 2009; see also 3.1):

$$x^{(i+1)} = \arg\min_{x} U\left(R_s^{(i)}(x)\right) \tag{10.1}$$

where $R_s^{(i)}$ is the surrogate model at iteration i. Here, $x^{(0)}$ is the initial design. $R_s^{(i)}$ is assumed to be a computationally cheap and sufficiently reliable representation of R_f, particularly in the neighborhood of $x^{(i)}$. Under these assumptions, the algorithm (10.1) is likely to produce a sequence of designs that quickly approach x_f^*. Usually, R_f is only evaluated once per iteration (at every new design $x^{(i+1)}$) for verification purposes and to obtain the data necessary to update the surrogate model. Because of the low computational cost of the surrogate model, its optimization cost can usually be neglected, and the total optimization cost is determined by the evaluation of R_f. The key point here is that the number of evaluations of R_f for a well-performing surrogate-based algorithm is substantially smaller than for most conventional optimization methods.

10.1.2 Robustness of the SBO Process

Robustness of the algorithm (10.1) depends on the quality of the surrogate model $R_s^{(i)}$. In general, in order to ensure convergence of the algorithm (10.1) to at least local optimum of the high-fidelity model, the first-order consistency conditions have to be met (Alexandrov and Lewis 2001), i.e., one has to have $R_s^{(i)}(x^{(i)}) = R_f(x^{(i)})$ and $JR_{s(i)}(x^{(i)}) = JR_f(x^{(i)})$, where J stands for the Jacobian of the respective model. Also, the process (10.1) has to be embedded in the trust-region (TR) framework (Conn et al. 2000), i.e., we have

$$x^{(i+1)} = \arg\min_{x:\ \|x-x^{(i)}\| \le \delta^{(i)}} U\left(R_s^{(i)}(x)\right) \tag{10.2}$$

where the TR radius $\delta^{(i)}$ is updated using classical rules (Conn et al. 2000). In general, the SBO algorithm (10.2) can be successfully utilized without satisfying the aforementioned conditions, see, e.g., Bandler et al. (2004a, b)and Koziel et al. (2008b). However, in these cases, the quality of the underlying low-fidelity model may be critical for performance (including the algorithm convergence) (Koziel et al. 2008a), and accurate location of the optimum design may not be possible.

Availability of cheap adjoint sensitivity (Nair and Webb 2003; CST MWS, 2013) makes it possible to satisfy consistency conditions in an easy way (without excessive computational cost by using, e.g., finite differentiation). A few options exploiting this possibility are discussed in the next sections.

10.2   SBO with First-Order Taylor Model and Trust Regions

The simplest way of exploring adjoint sensitivity for antenna optimization is to use the following surrogate model for the SBO scheme (10.2):

$$R_s^{(i)}(x) = R_f\left(x^{(i)}\right) + J_{R_f}\left(x^{(i)}\right)\cdot\left(x - x^{(i)}\right) \tag{10.3}$$

where J_{Rf} is the Jacobian of R_f obtained using adjoint sensitivity technique. The key point of the algorithm is finding the new design $x^{(i)}$ and the updating process for the search radius $\delta^{(i)}$. Here, instead of the standard rules, we use the following strategy ($x^{(i-1)}$ and $\delta^{(i-1)}$ are the previous design and the search radius, respectively):

1. For $\delta_k = k\cdot\delta^{(i-1)}$, $k=0, 1, 2$, solve $x^k = \arg\min\limits_{x:\|x-x^{(i)}\|\leq\delta_k} U\left(R_s^{(i)}(x)\right)$.

   Note that $x^0 = x^{(i-1)}$. The values of δ_k and $U_k = U(R_s^{(i)}(x^k))$ are interpolated using 2nd-order polynomial to find δ^* that gives the smallest (estimated) value of the specification error (δ^* is limited to $3\cdot\delta^{(i-1)}$). Set $\delta^{(i)} = \delta^*$.
2. Find a new design $x^{(i)}$ by solving 10.2 with the current $\delta^{(i)}$.
3. Calculate the gain ratio $\rho = [U(R_f(x^{(i)})) - U_0]/[U(R_s^{(i)}(x^{(i)})) - U_0]$; If $\rho < 0.25$, then $\delta^{(i)} = \delta^{(i)}/3$; else if $\rho > 0.75$ then $\delta^{(i)} = 2\cdot\delta^{(i)}$.
4. If $\rho < 0$ go to 2.
5. Return $x^{(i)}$ and $\delta^{(i)}$.

The trial points x^k are used to find the best value of the search radius, which is further updated based on the gain ratio ρ (actual versus expected objective function improvement). If the new design is worse than the previous one, the search radius is reduced to find $x^{(i)}$ again, which eventually will bring the improvement of U as $R_s^{(i)}$ and R_f are first-order consistent (Alexandrov and Lewis 2001). This precaution is necessary because the procedure in Step 1 only gives an estimation of the search radius.

Operation of the above algorithm is demonstrated below using two antenna designs: a planar inverted-F antenna (PIFA) and a wideband hybrid antenna. The

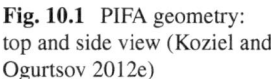
Fig. 10.1 PIFA geometry: top and side view (Koziel and Ogurtsov 2012e)

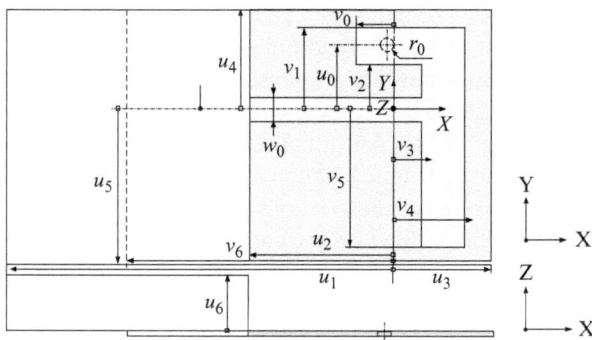

Table 10.1 PIFA: design results (Koziel and Ogurtsov 2012e)

Algorithm	Max$\lvert S_{11}\rvert$ for 1.82.1 GHz at the final design	Design cost (number of EM analyses)
Matlab's *fminimax*	−18.4 dB	105
This work	−17.5 dB	41

quality of the optimized design and computational cost of the process are compared to a benchmark technique, which is a Matlab's *fminimax* routine (Matlab 2012) which also uses adjoint sensitivity data.

10.2.1 Planar Inverted-F Antenna (PIFA)

Consider a PIFA (Volakis 2007) shown in Fig. 10.1 with 0.508 mm Rogers TMM4 substrate (Fig. 10.1, the upper panel) and Rogers TMM6 brick (Fig. 10.1, the lower panel). Fixed and dependent parameters are $[u_0\ u_1\ u_2\ u_3\ u_4\ u_5\ u_6\ u_7\ w_0\ r_0]^T = [6.15-50.0-15.0\ v_3+1\ 29.35\ 11.65\ 5.0\ 1.0\ 0.6]^T$ mm. The design goal is to adjust the geometry parameters so that $\lvert S_{11}\rvert \leq -15$ dB for 1.8–2.1 GHz. A requirement for the peak gain to be not less than 2.5 dBi for 1.8–2.1 GHz is implemented as a design constraint. The design variables are $\boldsymbol{x} = [v_0\ v_1\ v_2\ v_3\ v_4\ v_5\ v_6]^T$. The initial design is $\boldsymbol{x}^{(0)} = [-1.0\ 6.65\ 5.65\ 4.0\ 8.5\ -24.35\ -37.5]^T$ mm.

The final design with the algorithm of Sect. 10.2 is $\boldsymbol{x}^{(*)} = [-1.88\ 7.21\ 6.71\ 1.80\ 11.34\ -22.34\ -43.08]^T$ mm, and the one with Matlab's *fminimax* $\boldsymbol{x}^{(**)} = [-0.69\ 7.19\ 6.69\ 1.07\ 13.84\ -20.71\ -41.36]^T$ mm. Table 10.1 and Figs. 10.2, 10.3, and 10.4 compare the design cost and quality of the final design found by the SBO algorithm and Matlab's *fminimax*, which also uses adjoint sensitivity data in terms of the reflection and radiation responses: the algorithm of Sect. 10.2 allows 61 % reduction of the design cost in this example with only slight deterioration of the quality.

Fig. 10.2 PIFA: reflection at the initial design (*dotted line*) and at the final design with Matlab's *fminimax* (*dashed line*) and with the algorithm of Sect. 10.2 (*solid line*), (Koziel and Ogurtsov 2012e)

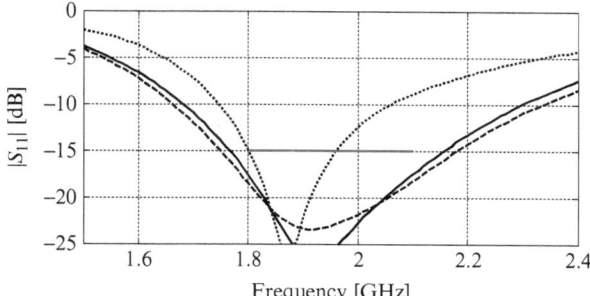

Fig. 10.3 PIFA: peak gain versus frequency at the initial design (*dotted line*), and the final design found by Matlab's *fminimax* (*dashed line*) and by the algorithm of Sect. 10.2 (*solid line*). The constraint level is shown with the *horizontal solid line* (Koziel and Ogurtsov 2012e)

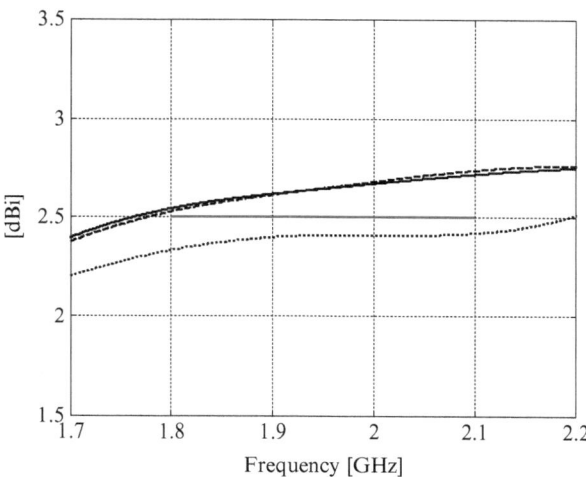

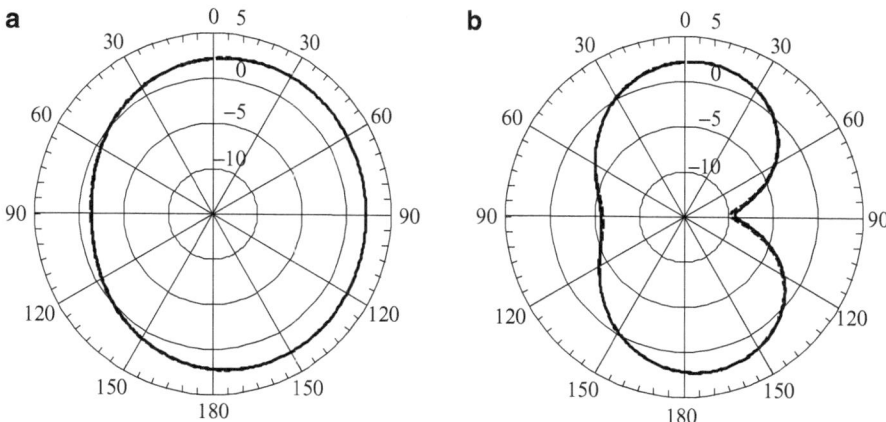

Fig. 10.4 PIFA: gain [dBi] at 1.95 GHz : (**a**) in the YOZ plane, 90° degree on the left corresponds to the positive Y-direction; (**b**) in the XOZ plane, 90° degree on the left corresponds to the positive X-direction. The gain patterns of the final designs found by Matlab's *fminimax* (*dashed line*) and by the algorithm of Sect. 10.2 (*solid line*) are hardly distinguishable (Koziel and Ogurtsov 2012e)

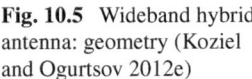

Fig. 10.5 Wideband hybrid antenna: geometry (Koziel and Ogurtsov 2012e)

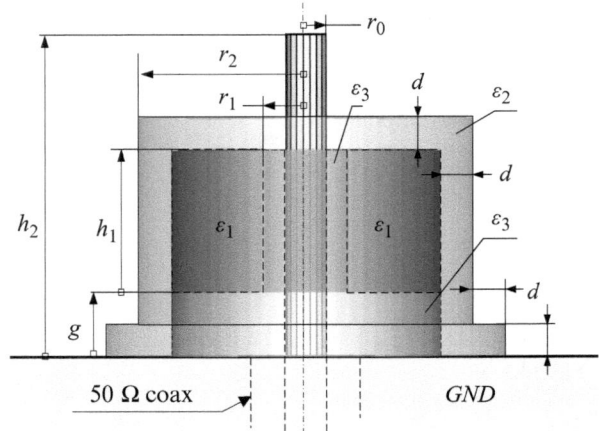

Table 10.2 Wideband hybrid antenna: design results (Koziel and Ogurtsov 2012e)

Algorithm	Max$|S_{11}|$ for 1.8–2.1 GHz at the final design	Design cost (number of EM analyses)
Matlab's *fminimax*	−22.6 dB	98
This work	−24.6 dB	24

10.2.2 Wideband Hybrid Antenna

The antenna shown in Fig. 10.5, a monopole loaded by a suspended dielectric ring resonator (DR) (Petosa 2007), should be optimized so that $|S_{11}| \leq -20$ dB for 8–13 GHz (Koziel and Ogurtsov 2012e). The design variables are $x = [h_1 h_2 r_1 r_2 g]^T$. The initial design is $x^{(0)} = [2.5\ 9.4\ 2.3\ 3.0\ 0.5]^T$mm. Fixed parameters are $[r_0\ d]^T = [0.635\ 1.00]^T$mm where r_0 is the radius of the probe and the inner conductor of the 50 Ω coax. The coax and the probe-dielectric resonator spacing are filled by Teflon which is also a material of the supporting ring of thickness g. Relative permittivity ε_1 and loss tangent of the DR are 10 and 0.001, respectively, at 10 GHz. Relative permittivity ε_2 and loss tangent of the housing are 2.7 and 0.01, respectively, at 10 GHz.

The final design with the algorithm of Sect. 10.2 is $x^{(*)} = [3.94\ 10.01\ 2.23\ 3.68\ 0.0]^T$mm and that with Matlab's *fminimax* is $x^{(**)} = [3.4986\ 9.9565\ 1.8489\ 3.9994\ 0.1267]$. Table 10.2 and Figs. 10.6 and 10.7 compare the design cost and quality of the final design found by the algorithm of Sect. 10.2 and Matlab's *fminimax*. Notice: both algorithms use adjoint sensitivity data.

It can be observed that the algorithm of Sect. 10.2 yields better design at significantly smaller computational cost (75% design time reduction). In addition, the final design of our algorithm is more preferable from an implementation point of view because it needs no supporting ring ($g = 0.0$ in $x^{(*)}$), while the final design of the benchmark algorithm $x^{(**)}$ resulted in thickness $g = 0.127$ mm.

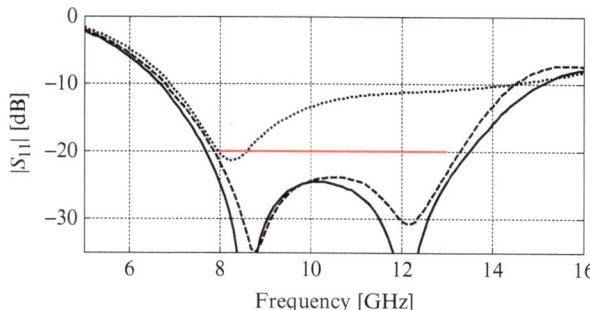

Fig. 10.6 Wideband hybrid antenna: reflection response at the initial design (*dotted line*), at the final design by Matlab's *fminimax* (*dashed line*), and by the algorithm of Sect. 10.2 (*solid line*), (Koziel and Ogurtsov 2012e)

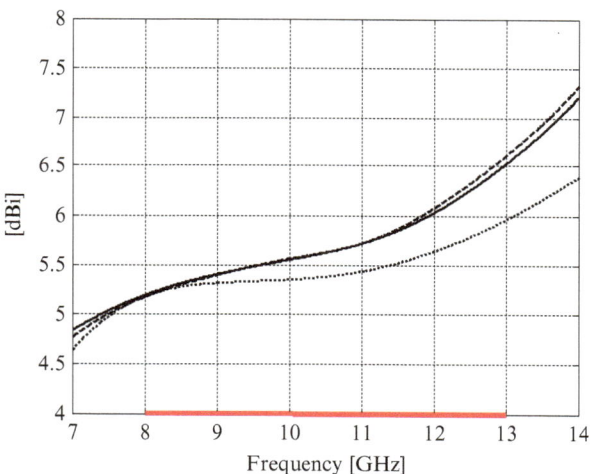

Fig. 10.7 Wideband hybrid antenna: realized peak gain versus frequency at the initial design (*dotted line*), and the final design found by Matlab's *fminimax* (*dashed line*) and by the algorithm of Sect. 10.2 (*solid line*). The optimization bandwidth is shown with the *horizontal solid line* (Koziel and Ogurtsov 2012e)

10.3 SBO with Space Mapping and Manifold Mapping

Construction of the surrogate model can be based on the underlying low-fidelity (or coarse) model R_c, e.g., obtained from coarse-discretization EM simulation data. The two methods considered here that use this approach are space mapping (SM) (Koziel et al. 2008a) and manifold mapping (MM) (Echeverria and Hemker 2005). Usually, the knowledge about the system embedded in the low-fidelity model allows us to reduce the number of high-fidelity model evaluations necessary to find an optimum design.

10.3.1 Surrogate Construction Using SM and Sensitivity Data

The space-mapping (SM) surrogate model is constructed using input and output SM (Bandler et al. 2004a, b) of the form

$$R_s^{(i)}(x) = R_c\left(x + c^{(i)}\right) + d^{(i)} + E^{(i)}\left(x - x^{(i)}\right) \tag{10.4}$$

Here, only the input SM vector $c^{(i)}$ is obtained through the nonlinear parameter extraction process

$$c^{(i)} = \arg\min_c \| R_f\left(x^{(i)}\right) - R_c\left(x^{(i)} + c\right) \| \tag{10.5}$$

Output SM parameters are calculated as

$$d^{(i)} = R_f\left(x^{(i)}\right) - R_c\left(x^{(i)} + c^{(i)}\right) \tag{10.6}$$

and

$$E^{(i)} = J_{R_f}\left(x^{(i)}\right) - J_{R_c}\left(x^{(i)} + c^{(i)}\right) \tag{10.7}$$

Formulations 10.4–10.7 ensure zero- and first-order consistency (Alexandrov and Lewis 2001) between the surrogate and the fine model.

10.3.2   Surrogate Construction Using MM and Sensitivity Data

The manifold-mapping (MM) surrogate model is defined as (Echeverria and Hemker 2005)

$$R_s^{(i)}(x) = R_f\left(x^{(i)}\right) + S^{(i)}\left(R_c(x) - R_c\left(x^{(i)}\right)\right) \tag{10.8}$$

where $S^{(i)}$ is the $m \times m$ correction matrix defined as

$$S^{(i)} = J_{R_f}\left(x^{(i)}\right) \cdot J_{R_c}\left(x^{(i)}\right)^{\dagger} \tag{10.9}$$

The pseudoinverse, denoted by †, is defined as

$$J_{R_c}^{\dagger} = V_{J_{R_c}} \Sigma_{J_{R_c}}^{\dagger} U_{J_{R_c}}^{T} \tag{10.10}$$

where UJR_c, ΣJR_c, and VJR_c are the factors in the singular value decomposition of JR_c. The matrix $\Sigma JR_c^{\dagger}$ is the result of inverting the nonzero entries in ΣJR_c, leaving the zeroes invariant (Echeverria and Hemker 2005). Using the sensitivity data as in 10.10 ensures that the surrogate model (10.8) is first-order consistent with the fine model. In our implementation, the coarse model is preconditioned using input space mapping of the form 10.5 in order to improve its initial alignment with the fine model.

Both the parameter extraction (10.5) and surrogate model optimization processes (10.2) are implemented by exploiting adjoint sensitivity data of the low-fidelity model, which allows for further cost savings. The details of these implementations can be found in Koziel et al. (2012c).

10.3.3 Fast Parameter Extraction and Surrogate Model Optimization

Sensitivity information can be utilized to speed up the parameter extraction process (10.5) as well as surrogate model optimization (10.1). In case of the parameter extraction process, we use a simple trust-region (Conn et al. 2000)-based algorithm, where the approximate solution $c^{(i,k+1)}$ of $c^{(i)}$ is found as (k is the iteration index for parameter extraction process (10.11))

$$c^{(i,k+1)} = \arg \min_{\|c-c^{(i,k)}\| \le \delta_{PE}^{(k)}} \| R_f\left(x^{(i)}\right) - L_{c.c}^{(i,k)}(c) \| \qquad (10.11)$$

where $L_{c.c}^{(i,k)}(c) = R_c(x^{(i)}+c^{(i,k)}) + JR_c(x^{(i)}+c^{(i,k)})\cdot(c-c^{(i,k)})$ is a linear approximation of $R_c(x^{(i)}+c)$ at $c^{(i,k)}$. The TR radius $\delta_{PE}^{(k)}$ is updated according to standard rules (Conn et al. 2000). Parameter extraction is terminated upon convergence or exceeding the maximum number of coarse model evaluations (in this work, the limit is set to 5 which is sufficient when using adjoint sensitivity).

Adjoint sensitivities are also utilized to lower the cost of surrogate model optimization. Similarly to 10.11, we use a TR-based algorithm that produces a sequence of approximations $x^{(i+1,k)}$ of the solution $x^{(i+1)}$ to 10.1 as follows (k is the iteration index for surrogate model optimization process (10.12)):

$$x^{(i+1,k+1)} = \arg \min_{\|x-x^{(i+1,k)}\| \le \delta_{SO}^{(k)}} U\left(L_{c.x}^{(i,k)}(x)\right) \qquad (10.12)$$

where $L_{c.x}^{(i,k)}(x) = R_s^{(i)}(x^{(i+1,k)}+c^{(i)}) + JR_{s(i)}(x^{(i+1,k)}+c^{(i)})\cdot(x-x^{(i+1,k)})$ is a linear approximation of $R_s^{(i)}(x+c^{(i)})$ at $x^{(i+1,k)}$. The TR radius $\delta_{SO}^{(k)}$ is updated according to standard rules (Koziel et al. 2010d). Typically, due to adjoint sensitivities, surrogate model optimization requires only a few evaluations of the coarse model R_c. Note that sensitivities of the surrogate model can be calculated using the sensitivities of both R_f and R_c as follows: $JR_{s(i)}(x+c^{(i)}) = JR_c(x+c^{(i)}) + [JR_f(x^{(i)}) - JR_c(x^{(i)}+c^{(i)})]$.

10.3.4 UWB Monopole Optimization Using SM and MM Surrogates

Consider the UWB antenna shown in Fig. 10.8. The antenna models include microstrip monopole, housing, edge mount SMA connector, and section of the feeding coax.

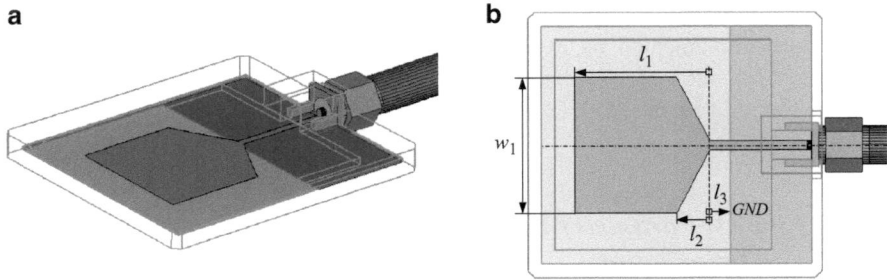

Fig. 10.8 UWB monopole: (**a**) 3D view and (**b**) top view with substrate and housing shown transparent (Koziel and Ogurtsov 2012d)

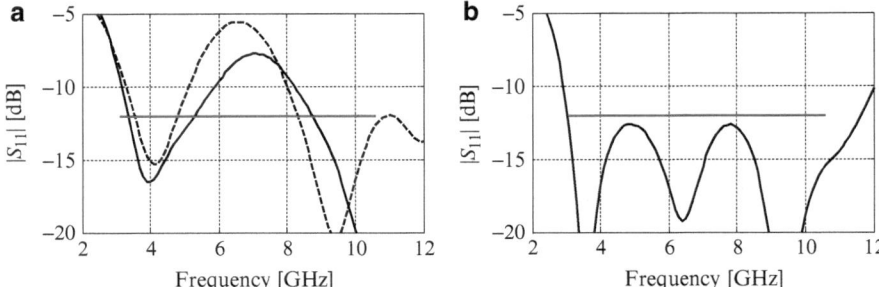

Fig. 10.9 UWB monopole optimized using the MM algorithm: (**a**) responses of R_f (*solid line*) and R_c (*dashed line*) at the initial design x^{init}; (**b**) response of R_f (*solid line*) at the final design (Koziel and Ogurtsov 2012d)

The design variables are $x = [l_1 \, l_2 \, l_3 \, w_1]^T$. The initial design is $x^{init} = [20 \, 2 \, 0 \, 25]^T$ mm. Simulation time of the low-fidelity model R_c (156,000 mesh cells) is 1 min and that of the high-fidelity model R_f (1,992,060 mesh cells) is 40 min (both at the initial design). Both models are simulated with the transient solver of CST Microwave Studio (CST MWS 2013). The design specifications are $|S_{11}| \leq -12$ dB for 3.1 GHz to 10.6 GHz.

The monopole was optimized using the SBO algorithm (10.2) with both the SM and MM surrogate models. Figure 10.9a shows the responses of R_f and R_c at x^{init}. Figure 10.9b shows the response of the high-fidelity model at the final design $x^{(2)} = [20.22 \, 2.43 \, 0.128 \, 19.48]^T$ ($|S_{11}| \leq -12.5$ dB for 3.1 to 10.6 GHz) obtained after two SBO iterations with the MM surrogate, i.e., only 4 evaluations of the high-fidelity model R_f (Table 10.3). Figure 10.10 shows the evolution of the specification error with the MM algorithm. The number of function evaluations is larger than the number of MM iterations because some designs can be rejected by the TR mechanism. The algorithm using the SM surrogate required three iterations, and the final design is $x^{(3)} = [20.29 \, 2.27 \, 0.058 \, 19.63]^T$ ($|S_{11}| \leq -12.8$ dB for 3.1 to 10.6 GHz) obtained after three SM iterations. The total optimization cost (Table 10.4) is equivalent to around 6 evaluations of the fine model.

Table 10.3 UWB monopole with manifold mapping ($|S_{11}| \leq -12.5$ dB, $3.1 - 10.6$ GHz): optimization costs (Koziel and Ogurtsov 2012d)

Algorithm component	Number of model evaluations[a]	CPU time Absolute	Relative to $\boldsymbol{R}_f$
Evaluation of $\boldsymbol{R}_c$	31	31 min	0.8
Evaluation of $\boldsymbol{R}_f$	4	120 min	4.0
Total cost[a]	N/A	151 min	**4.8**

[a]Includes $\boldsymbol{R}_f$ evaluation at the initial design

Fig. 10.10 UWB monopole: minimax specification error versus iteration index for the SBO algorithm using the MM surrogate model (Koziel and Ogurtsov 2012d)

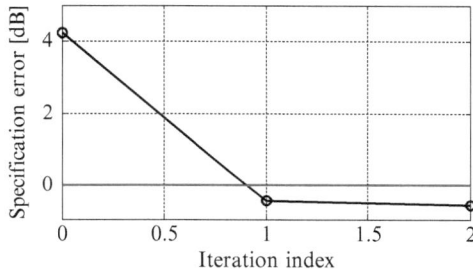

Table 10.4 UWB monopole with space mapping ($|S_{11}| \leq -12.8$ dB, $3.1 - 10.6$ GHz): optimization costs (Koziel and Ogurtsov 2012d)

Algorithm component	Number of model evaluations[a]	CPU time Absolute	Relative to $\boldsymbol{R}_f$
Evaluation of $\boldsymbol{R}_c$	45	45 min	1.1
Evaluation of $\boldsymbol{R}_f$	5	200 min	5.0
Total cost[a]	N/A	205 min	**6.1**

[a]Includes $\boldsymbol{R}_f$ evaluation at the initial design

10.4   SPRP with Adjoint Sensitivity

In a way similar to that presented in Sect. 10.2, sensitivity information can be utilized to enhance most surrogate models. More specifically, the sensitivity-enhanced model may be defined as (Koziel and Ogurtsov 2013f)

$$\boldsymbol{R}_{s.sens}^{(i)}\left(\boldsymbol{x}\right) = \boldsymbol{R}_s^{(i)}\left(\boldsymbol{x}\right) + \left[\boldsymbol{J}_{\boldsymbol{R}_s^{(i)}}\left(\boldsymbol{x}^{(i)}\right) - \boldsymbol{J}_{\boldsymbol{R}_f}\left(\boldsymbol{x}^{(i)}\right)\right] \cdot \left(\boldsymbol{x} - \boldsymbol{x}^{(i)}\right) \qquad (10.13)$$

where $\boldsymbol{R}_s^{(i)}$ is the standard surrogate, whereas $\boldsymbol{JR}_{s(i)}$ and $\boldsymbol{JR}_f$ are Jacobians of the standard surrogate and the high-fidelity models, respectively. The Jacobian of the high-fidelity model is obtained directly from the EM solver.

Here, we consider sensitivity enhancement of the SPRP surrogate (cf. Sect. 4.3). The Jacobian of $\boldsymbol{R}_s^{(i)}$ is obtained indirectly using the Jacobian of the low-fidelity

Table 10.5 UWB monopole with the adjoint sensitivity-enhanced SPRP algorithm ($|S_{11}| \leq -13.3$ dB, $3.1 - 10.6$ GHz): optimization costs (Koziel and Ogurtsov 2013f)

Algorithm	Algorithm component	Model evaluations[a]	CPU time Absolute(min)	Relative to R_f
SPRP	Evaluation of R_c	**34** [88][b]	**34** [88]	**0.8** [2.2]
	Evaluation of R_f	**5** [3]	**200** [120]	**5.0** [3.0]
	Total cost[a]	N/A	**234** [208]	**5.8** [5.2]
Matlab's *fminimax*	Total cost	$62 \times R_f$	2,480	62.0

[a]Excludes R_f evaluation at the initial design
[b]Numbers in brackets correspond to the original SPRP algorithm not using sensitivity data

model and finite differentiation at the SPRP characteristic points. The standard SPRP surrogate ensures a zero-order consistency condition of $R_s^{(i)}(x^{(i)}) = R_f(x^{(i)})$. The enhanced model (10.13) also ensures first-order consistency, i.e., $JR_{s.sens(i)}(x^{(i)}) = JR_f(x^{(i)})$, which improves local search capability of the SPRP optimization process. Moreover, the search process is embedded in the trust-region framework (10.2). Sensitivity information is also utilized to speed up the SPRP optimization step (10.2) as described in Sect. 10.3.3.

The enhanced SPRP technique was used to optimize the UWB monopole, Fig. 10.8, with the same design specifications and initial design as those of Sect. 10.3.1. The enhanced SPRP technique produced the final design $x^* = [20.14 \ 19.02 \ 2.30 \ 0.0396]^T$ mm with $|S_{11}| \leq -13.3$ dB for $3.1 - 10.6$ GHz. The original SPRP technique produced the final design with $|S_{11}| \leq -12.1$ dB for $3.1 - 10.6$ GHz. The computational costs of both SPRP versions are similar and listed in Table 10.5; however, the SPRP algorithm enhanced with adjoint sensitivity produced the best design. It can be seen that adjoint sensitivity enhancement is crucial here: the original SPRP algorithms stops after 2 iterations.

10.5   Discussion and Conclusion

Cheap derivative information obtained using adjoint sensitivity technique may be useful to improve the robustness and computational complexity of simulation-driven design of antenna structures. As demonstrated in Sect. 10.2, trust-region-based algorithms using first-order Taylor models can be more efficient than state-of-the-art general purpose gradient-based algorithms, which is because the latter are capable of handling difficult situations (e.g., involved nonlinear constraints), which is usually not necessary for antenna problems where simple lower/upper bounds for parameters as well as linear constraints are normally sufficient. On the other hand, sensitivity data may also be useful to enhance variable-fidelity algorithms such as space mapping, manifold mapping, or shape-preserving response prediction. In those cases, sensitivity can help not only in better alignment between the surrogate and the high-fidelity EM model to be optimized but also in reducing the cost of optimizing the surrogate model itself.

Chapter 11
Simulation-Based Multi-objective Antenna Optimization with Surrogate Models

Practical antenna design is a multi-objective task. In many situations, it is possible to identify the most important objective (e.g., reflection requirements) and handle antenna design using conventional, single-objective algorithms. This approach was used in the previous chapters, where secondary objectives (e.g., gain performance requirements) were handled through design constraints, either as explicit constraints or using penalty functions. In general, however, having more than one objective substantially complicates the design process because if the designer priorities are not clearly defined beforehand, multi-objective optimization becomes a necessity (Koulouridis et al. 2007; Kuwahara 2005). The goal of formal multi-objective optimization is to find the so-called Pareto front representing the best possible trade-offs between conflicting objectives. Probably the most popular approaches to solve multi-objective antenna design problems are metaheuristic (or population-based) algorithms such as genetic algorithms (GAs) and particle swarm optimizers (PSO) (e.g., Koulouridis et al. 2007; Jin and Rahmat-Samii 2007). The advantage of metaheuristics is their ability to find the entire Pareto front in one algorithm run. The drawback is a large number (hundreds, thousands, or even tens of thousands) of objective function evaluations required, which often turns in a prohibitive computational time if the objective functions are supplied by full-wave discrete simulators.

In this chapter, we present a multi-objective design procedure for antennas that allows us to obtain a Pareto front, i.e., multiple designs representing the trade-off between various characteristics of the antenna under consideration. In order to reduce the computational cost of the design process, we exploit a fast surrogate model constructed by approximating coarse-discretization EM simulation of the antenna structure. The surrogate is optimized with respect to the objectives of interest using a multi-objective evolutionary algorithm. The selected elements of the Pareto front obtained this way are further refined using high-fidelity EM simulations and surrogate-based optimization (Koziel 2011) exploiting an appropriately corrected initial surrogate. This allows us to obtain a set of high-fidelity Pareto optimal designs at a very low CPU cost.

S. Koziel and S. Ogurtsov, *Antenna Design by Simulation-Driven Optimization*,
SpringerBriefs in Optimization, DOI 10.1007/978-3-319-04367-8_11,
© Slawomir Koziel and Stanislav Ogurtsov 2014

Our technique is illustrated using two antenna examples each having two objectives: a UWB monopole antenna with minimization of the antenna reflection and antenna size and a planar Yagi antenna with minimization of the antenna reflection and maximization of the antenna end-fire gain over certain frequency bands.

11.1   Multi-objective Antenna Design Using Surrogate Modeling and Evolutionary Algorithms

In this section, we describe a multi-objective surrogate-based optimization procedure. We begin with formulating the multi-objective antenna design problem. Subsequently, we discuss the optimization approach. The illustration examples are presented in Sects. 11.2 and 11.3.

11.1.1   Multi-objective Antenna Design Problem

Let $R_f(x)$ be a response of an accurate model of the antenna under consideration. The response $R_f(x)$ is computed using high-fidelity EM simulation, and it may represent an antenna reflection coefficient. Here, x is a vector of designable parameters, i.e., antenna dimensions.

Let $F_k(x)$, $k = 1, ..., N_{obj}$, be a kth design objective. A typical performance objective would be to minimize antenna reflection over a certain frequency band of interest and to ensure that $|S_{11}| < -10$ dB over that band. There might be also geometrical objectives such as to minimize $F_k(x) = A(x)$ – the antenna size defined in a convenient way (e.g., maximal lateral size, height, the maximal dimension, area of the footprint, antenna volume). Similar objectives can be formulated with respect to antenna gain, radiation pattern, efficiency, etc., e.g., to minimize one over the broadside (or end-fire) gain over the frequencies of interest.

If $N_{obj} > 1$ then any two designs $x^{(1)}$ and $x^{(2)}$ for which $F_k(x^{(1)}) < F_k(x^{(2)})$ and $F_l(x^{(2)}) < F_l(x^{(1)})$ for at least one pair $k \neq l$ are not commensurable, i.e., none is better than the other in the multi-objective sense. We define Pareto dominance relation $\prec$ (Fonseca 1995) saying that for the two designs x and y, we have $x \prec y$ (x dominates y) if $F_k(x) < F_k(y)$ for all $k = 1, ..., N_{obj}$. The goal of multi-objective optimization is to find a representation of a so-called Pareto front (or Pareto optimal set) X_P of the design space X, such that for any $x \in X_P$, there is no $y \in X$ for which $y \prec x$ (Fonseca 1995).

11.1.2   Optimization Algorithm

Because the high-fidelity model R_f is computationally too expensive to be directly handled in multi-objective optimization, we use a surrogate model constructed as follows. Let $\boldsymbol{R}_{cd}$ be a coarse-discretization EM simulation model of the antenna. Typically, $\boldsymbol{R}_{cd}$ is 10–50 times faster than $\boldsymbol{R}_f$ but it is still too expensive for multi-objective optimization. Therefore, we sample the design space using Latin hyper-cube sampling (Beachkofski et al. 2002) and create a fast surrogate model $\boldsymbol{R}_s$ by approximating the sampled $\boldsymbol{R}_{cd}$ data using kriging interpolation (Queipo et al 2005). The kriging model $\boldsymbol{R}_s$ is very fast, smooth, and easy to optimize. In the next stage, we apply a multi-objective evolutionary algorithm (MOEA) to optimize $\boldsymbol{R}_s$ and to find a set of designs representing Pareto optimal solutions with respect to the objectives F_k of interest. Here, we use a standard multi-objective evolutionary algorithm with fitness sharing, Pareto dominance tournament selection, and mating restrictions (Fonseca 1995).

Let $\boldsymbol{x}_s^{(k)}$, $k = 1, \ldots, K$, be the selected elements of the Pareto front found by the MOEA. These solutions have to be refined because they were obtained by optimizing the surrogate model, whereas we are interested in optimizing the high-fidelity model. For simplicity of the notation, the design refinement stage below is defined assuming two objectives F_1 and F_2; however, it can be generalized for any value of N_{obj}. The refinement stage exploits the output space-mapping (OSM) (Koziel et al. 2008b) process:

$$\boldsymbol{x}_f^{(k \cdot i + 1)} = \arg \min_{x, F_2(x) \leq F_2\left(x_s^{(k \cdot i)}\right)} F_1\left(\boldsymbol{R}_s(x) + \left[\boldsymbol{R}_f\left(\boldsymbol{x}_s^{(k \cdot i)}\right) - \boldsymbol{R}_s\left(\boldsymbol{x}_s^{(k \cdot i)}\right)\right]\right) \quad (11.1)$$

The optimization process (11.1) is constrained not to increase the second objective as compared to $\boldsymbol{x}_s^{(k)}$. The surrogate model $\boldsymbol{R}_s$ is corrected using the OSM term $\boldsymbol{R}_f(\boldsymbol{x}_s^{(k.i)}) - \boldsymbol{R}_s(\boldsymbol{x}_s^{(k.i)})$ (here, $\boldsymbol{x}_f^{(k.0)} = \boldsymbol{x}_s^{(k)}$) so that the corrected surrogate model coincides with $\boldsymbol{R}_f$ at the beginning of each iteration. In practice, two or three iterations of 11.1 are sufficient to find a refined high-fidelity model design $\boldsymbol{x}_f^{(k)}$. After completing this stage, we create a set of Pareto optimal high-fidelity model designs. This set is the final outcome of our multi-objective optimization process.

The entire design flow can be summarized as follows:

1. Sample the design space and acquire the $\boldsymbol{R}_{cd}$ data.
2. Construct the kriging interpolation model $\boldsymbol{R}_s$.
3. (Optional) Correct the kriging model $\boldsymbol{R}_s$ using space mapping.
4. Obtain the Pareto front by optimizing $\boldsymbol{R}_s$ using MOEA.
5. Refine selected elements of the Pareto front, $\boldsymbol{x}_s^{(k)}$, to obtain corresponding high-fidelity model designs $\boldsymbol{x}_f^{(k)}$.

It should be emphasized that the high-fidelity model $\boldsymbol{R}_f$ is not evaluated until the refinement stage (step 5 above). It is also worth mentioning that finding the high-fidelity model Pareto optimal set requires only about three evaluations of the high-fidelity model per design. The optional step 3 can be executed in case of considerable discrepancy between $\boldsymbol{R}_s$ and $\boldsymbol{R}_f$. In that case, before finding the Pareto front, the kriging model is enhanced by aligning it with the high-fidelity model at certain (usually small) number of designs using space mapping. Typically, output space mapping and frequency scaling are preferred (cf. Sect. 4.2.3 for more details).

11.2   Application: A UWB Monopole

In this section, we demonstrate the operation of the multi-objective optimization algorithm for a UWB monopole with minimization of antenna reflection coefficient and size being the two objectives.

11.2.1   UWB Monopole: Geometry and Problem Statement

Consider a UWB monopole shown in Fig. 11.1. The antenna is energized through a 50 Ω coaxial input (Teflon filling, $r_0 = 0.635$ mm). No extra circuitry is used for matching here. Design specification imposed on the reflection response of the monocone is $|S_{11}| \leq -10$ dB from 3 to 10 GHz. Design variables are $x = [z_1 \ z_2 \ r_1]^T$ (sizes in mm), where z_1 is the extension of the coax pin, z_2 is the length of the cone section, and r_1 is the size of the radial line section as shown in Fig. 11.1b. The ground plane is modeled with infinite lateral extends.

The high-fidelity model of the antenna is simulated in CST Microwave Studio (CST MWS 2013) (~1,400,000 mesh cells, evaluation time 23 min). The coarse-discretization model R_{cd} is also simulated in CST MWS (~33,000 mesh cells, 33 s). The design space is defined by $0 \leq z_1 \leq 4$, $2 \leq z_2 \leq 15$, $4 \leq r_1 \leq 20$, and a linear constraint $z_1 + z_2 \leq r_1 - 0.25$. The antenna size defined here is the maximal dimension out of vertical and lateral ones: $A(x) = \max\{2r_2, z_1 + z_2 + r_2\}$, where $r_2 = (r_1^2 - (z_1 + z_2)^2)^{1/2}$ is the radius of the hemisphere terminating the conical section.

The design objectives for this example are the following:

1. $F_1(x) = \max\{|S_{11}(x,f)|:\ 3\ \text{GHz} \leq f \leq 10\ \text{GHz}\} - $ maximum of $|S_{11}|$ over the frequency band of 3–10 GHz.
2. $F_2(x) = A(x) - $ antenna size as defined in the previous paragraph.

11.2.2   UWB Monopole: Results

The kriging surrogate model R_s is created using 600 low-fidelity model samples allocated in the design space using Latin hypercube sampling (Beachkofski and

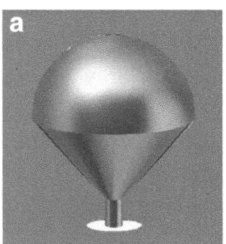
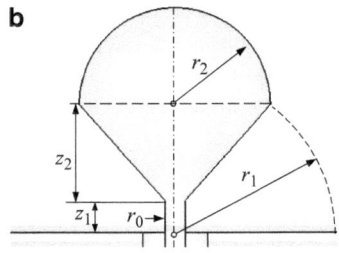

Fig. 11.1 UWB monopole: (**a**) 3D view and (**b**) cut view

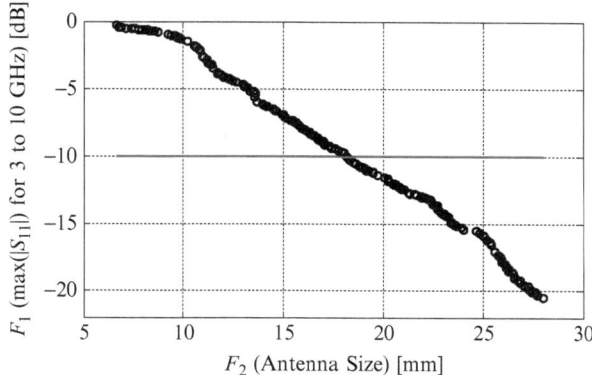

Fig. 11.2 Pareto front for the UWB monopole obtained by optimizing the kriging surrogate model using the multi-objective evolutionary algorithm

Table 11.1 UWB monopole: optimization results

Antenna size $A(x)$ (mm)	Design variables (mm)			Max $|S_{11}|$ [dB] for 3–10 GHz
	z_1	z_2	r_1	
18	0.994	10.36	13.16	−9.5
19	0.021	12.18	12.97	−10.9
20	0.065	12.48	14.59	−11.6
21	0.051	12.84	15.23	−12.3
22	0.000	12.92	15.79	−13.2
23	0.000	12.06	16.28	−14.7
24	0.008	12.08	16.97	−16.0
25	0.079	12.42	17.68	−17.1
26	0.142	12.99	18.38	−18.1
27	0.169	13.43	19.09	−18.4
28	0.231	13.27	19.45	−19.4

Grandhi 2002). The surrogate was then optimized using the multi-objective evolutionary algorithm (MOEA). Figure 11.2 shows the Pareto front obtained by applying the MOEA. It indicates that the minimum size of the considered antenna that ensures satisfying reflection requirements ($|S_{11}| \leq -10$ dB) is 18.5 mm.

A number of selected solutions of the Pareto front, corresponding to sizes of 18, 19, ..., 28, have been refined using the OSM algorithm (11.1) to obtain high-fidelity model designs. Those optimum solutions are gathered in Table 11.1.

The computational cost of creating the kriging surrogate corresponds to only about 15 evaluations of the high-fidelity model R_f. Multi-objective optimization of R_s takes about 10 min of CPU time as the kriging model is very fast. Design refinement takes, on average, two iterations of the OSM algorithm (i.e., two evaluations of R_f). Thus, the total design cost (obtaining Pareto front and 11 antenna designs of various sizes as in Table 11.1) corresponds to about 38 evaluations of the high-fidelity model (~ 14 h).

Figure 11.3 shows the Pareto optimal set of the high-fidelity model designs obtained after the design refinement. Figure 11.4 shows the reflection responses for

Fig. 11.3 Pareto optimal designs obtained in the refinement stage (cf. Table 11.1)

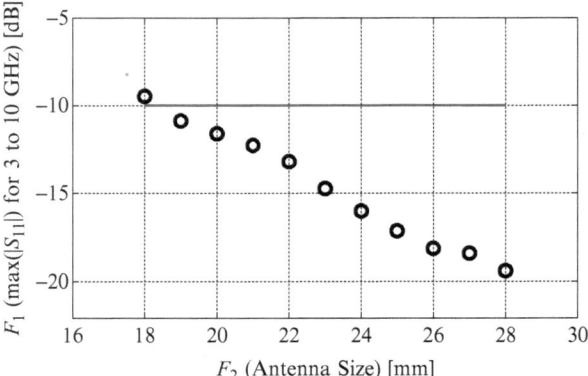

Fig. 11.4 Reflection responses of the UWB monopole: the optimized high-fidelity model designs of sizes 19 mm (*dotted line*), 22 mm (*dashed dotted line*), 24 mm (*dashed line*), and 28 mm (*solid line*)

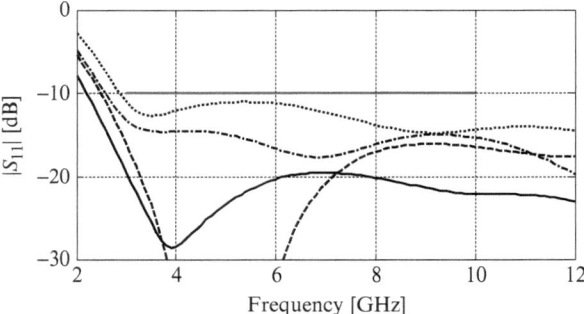

selected designs of this set. The monopole configuration is typically required to have omnidirectional broadside radiation. Figure 11.5 shows the simulated realized gain responses for selected designs of this set comparing their performance on how the designs provide the broadside performance. The 19 mm size design, having the highest reflection response in Fig. 11.4, definitely shows the broadside performance in Fig. 11.5a: its realized peak gain and maximum realized gain within the 30° elevation sector from the horizon overlap over the frequency band, i.e., the peak of radiation is contained within the sector at all frequencies, and moreover, it is at the zero elevation angle. The 22 mm size design is essentially of a broadside behavior showing a small difference between its realized peak gain and maximum realized gain within the 30° sector above 8 GHz in Fig. 11.5b. This small difference, nevertheless, indicates that the peak of radiation is not within the elevation sector after 8 GHz. Also the small difference between the maximum realized gain and the average realized gain for the 19 mm and the 22 mm designs, which is less than 1 dB for the former and up to 1.3 dB for the later, illustrates that the broadside radiation pattern of the two designs are quite uniform within the elevation sector. The 24 and 28 mm designs show more nonuniform radiation within the 30° elevation sector and over the frequency in Fig. 11.5c, d. The direction of their peak radiation is not within the angular sector in the 5.5–8.7 GHz band for the 24 mm design and in the

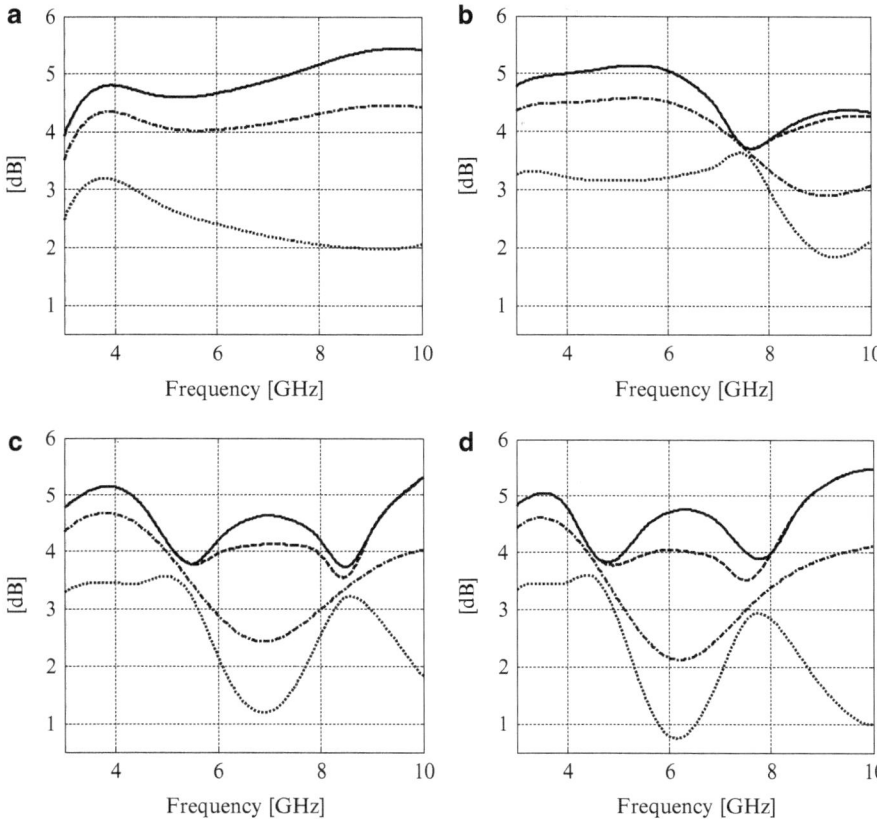

Fig. 11.5 Radiation responses of the UWB monopole corresponding to the optimized high-fidelity model designs of sizes (**a**) 19 mm, (**b**) 22 mm, (**c**) 24 mm, and (**d**) 28 mm: realized peak gain (*solid line*); maximum realized gain within the 30° elevation sector from the horizon, the ground plane of the model (*dashed line*); minimum realized gain within the 30° elevation sector (*dotted line*), average realized gain within the 30° elevation sector (*dashed dotted line*)

5–8 GHz band for the 28 mm design; also the differences between the average realized gain and maximum gain are higher than those of the 19 and the 22 mm designs after 3 GHz. At the same time, in terms of maximal and minimal values of realized gain or in terms of swings of the realized gains, all designs are quite close.

It should be emphasized that the approach described here allows, at a low computational cost, to obtain comprehensive information about a given UWB antenna, including the trade-offs between the reflection coefficient and the antenna size, as well as the minimum size while satisfying given reflection requirements. For the specific example considered here, the minimum size of the antenna still satisfying reflection requirements is 34 % smaller than the size of the antenna optimized for reflection only. It is also important that the total cost of obtaining the set of Pareto optimal high-fidelity model designs (here, less than 40 evaluations of R_f) is comparable to the cost of conventional single-objective optimization of the antenna structure yielding a single design.

11.3 Application: A Planar Yagi Antenna

In this section we demonstrate the operation of the multi-objective design methodology of Sect. 11.1 for a planar Yagi antenna (Deal et al. 2000; Kaneda et al. 2002) shown in Fig. 11.6. Here, the objectives are minimization of antenna reflection coefficient and maximization of end-fire gain.

11.3.1 Planar Yagi Antenna: Geometry, Models, and Problem Statement

The Yagi antenna of interest (layout shown in Fig. 11.6) comprises a driven element fed by a coplanar strip line, director, and microstrip balun. The substrate is a 0.635 mm thick Rogers RT6010 (CSR MWS 2013). Metallization is with 1 oz copper cladding. The antenna is fed with 50 Ω microstrip. Design variables are $x = [s_1\ s_2\ v_1\ v_2\ u_1\ u_2\ u_3\ u_4]^T$. Other dimension parameters are fixed as follows: $w_1 = w_3 = w_4 = 0.6$, $w_2 = 1.2$, $u_5 = 1.5$, $s_3 = 3.0$, and $v_3 = 17.5$, all in mm. The antenna substrate/ground is modeled to be of infinite lateral extend at the feed side.

The multi-objective design methodology of Sect. 11.1 cannot be applied directly for this case because of the large number of design variables. More specifically, the computational cost of creating the global and accurate kriging model of the antenna structure would be too high. Therefore, we decompose the structure into two parts as shown in Fig. 11.7: the antenna, Fig. 11.7b, and the balun, Fig. 11.7c. For the sake of building the surrogate model as described in Sect. 11.3.2, the antenna part, Fig. 11.7b, is described with the coarse-discretization model $R_{c,a}$ which contains only parts of the design variables, $x_a = [s_1\ s_2\ v_1\ v_2]^T$. The balun part, Fig. 11.7c, is described with the coarse-discretization model $R_{c,b}$ which contains the rest of the variables $x_b = [u_1\ u_2\ u_3\ u_4]^T$. The high-fidelity model R_f contains the complete set of the variables $x = [x_a\ x_b]^T$.

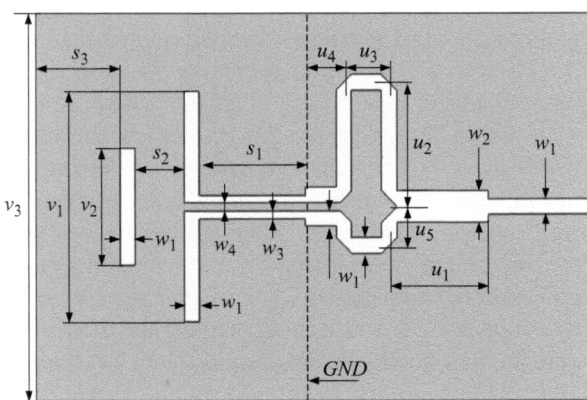

Fig. 11.6 Planar Yagi antenna: layout

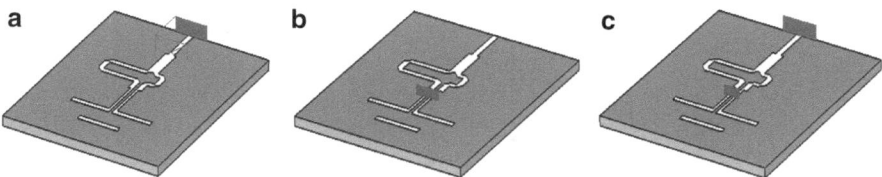

Fig. 11.7 EM models with excitation ports shown: (**a**) high-fidelity model R_f, (**b**) coarse-discretization model with excitation applied directly to coplanar line of the driven element $R_{c,a}$, (**c**) coarse-discretization model of the balun $R_{c,b}$ with two ports shown

Each coarse-discretization model is defined with its complement being inactive, as shown in Fig. 11.7b, c, using the internal excitation which is available with the transient solver of CST MWS (CST MWS 2013).

While the high-fidelity model is discretized with fine meshes, accounts for material losses, the coarse models are discretized with coarser meshes, assume lossless materials. In addition, the coarse models run with much more relaxed residual energy termination condition (-25 dB for $R_{c,a}$ and -25 dB for $R_{c,b}$). The high-fidelity model R_f contains 1,374,160 hexahedral mesh cells at the initial design and is simulated in 35 min 17 s. The coarse model $R_{c,a}$ contains 84,490 hexahedral mesh cells at the initial design and is simulated in 76 s. The coarse model $R_{c,b}$ contains 85,630 hexahedral mesh cells at the initial design and is simulated in 134 s. The low-fidelity model of the entire structure is set as follows: the radiation response is configured from that of model $R_{c,a}$ (shown in Fig. 11.8b) as an RSA surrogate model and the reflection response at a particular frequency point is obtained as

$$S_{11} = S_{11,b} + \frac{S_{12,b}S_{21,b}S_{11,a}}{1 - S_{22,b}S_{11,a}} \tag{11.2}$$

where $S_{ij,b}$ are the S-parameters of the surrogate model $R_{s,b}$ and $S_{11,a}$ is the reflection coefficient of the surrogate model $R_{s,a}$.

For the Yagi antenna considered here, we are going to study the antenna reflection and radiation responses using the MOEA and targeting the best matching and the highest average end-fire gain within the 10–11 GHz bandwidth.

11.3.2   Planar Yagi Antenna: Surrogate Models

The surrogate model R_s of the Yagi antenna for the purpose of finding the Pareto front (cf. Sect. 11.1.2) is constructed as follows. In the first stage, we sample and obtain the kriging model $R_{s,a}$ of the coarse-discretization model of the antenna proper $R_{c,a}$. The design space is defined as $3.8\,\text{mm} \le s_1 \le 4.4\,\text{mm}, 2.8\,\text{mm} \le s_2 \le 4.4\,\text{mm}$, $8.0\,\text{mm} \le v_1 \le 9.8\,\text{mm}$, and $4.0\,\text{mm} \le v_2 \le 5.2\,\text{mm}$. The kriging model is constructed using 256 samples allocated on a uniform rectangular grid.

Fig. 11.8 Model responses
at the initial design: (**a**)
reflection with R_f (*solid line*)
and that obtained from
S-parameters of the coarse
models $R_{c,a}$ and $R_{c,b}$; (**b**)
end-fire gain of R_f (*solid
line*), peak gain of R_f (*dotted
line*), end-fire gain of $R_{c,a}$
(*thick solid line*), peak gain of
the coarse-discretization
model of the whole antenna
with material losses neglected
(*dotted dashed line*), and
end-fire gain of the coarse-
discretization model of the
entire antenna with material
losses neglected (*dashed line*)

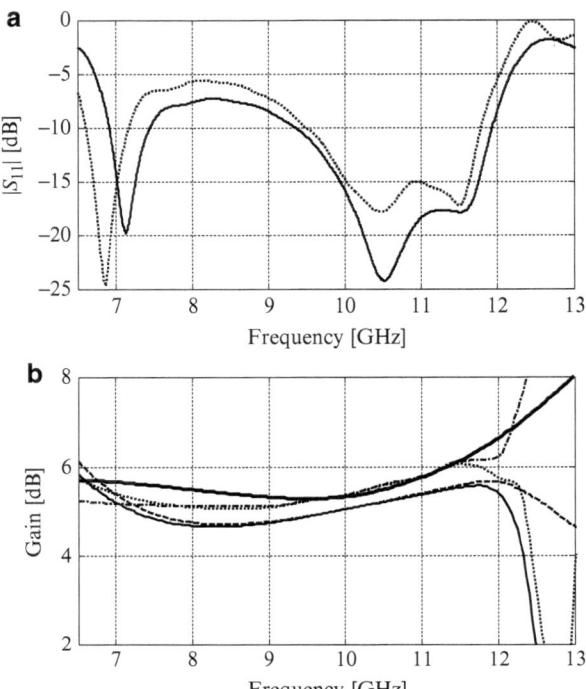

In the second stage, the kriging model $R_{s,b}$ of the balun is obtained. Here, the design space is defined as 3.0 mm $\leq u_1 \leq 4.2$ mm, 4.5 mm $\leq u_2 \leq 5.2$ mm, 1.8 mm $\leq u_3 \leq 2.6$ mm, and 1.3 mm $\leq u_4 \leq 1.8$ mm. The kriging model is constructed using 625 samples allocated on a uniform rectangular grid.

The initial surrogate model is obtained from the two aforementioned kriging models so that the antenna gain is directly modeled by $R_{s,a}$, whereas its reflection response is obtained from both $R_{s,a}$ and $R_{s,b}$ using 11.2.

Because of the discrepancy between R_s created as above and the high-fidelity model, it is further enhanced using space-mapping alignment with $2n+1$ (*n* being the number of design variables, here 8) high-fidelity model training samples allocated using the so-called star distribution (Bandler et al. 2004a, b). The space mapping used here is parameter shift (or input space mapping, cf. 3.13) as well as response correction (cf. Sect. 4.2.3). It should be reiterated that the major reason for decomposing the structure into two parts is that it allows to alleviate the curse of dimensionality problem, i.e., creating an accurate response surface model of the entire structure would require thousands of training samples. The total cost of creating the surrogate, including the space-mapping correction, corresponds to about 60 evaluations of the high-fidelity model.

Fig. 11.9 Pareto front for the planar Yagi antenna obtained by optimizing the RSA surrogate model using MOEA

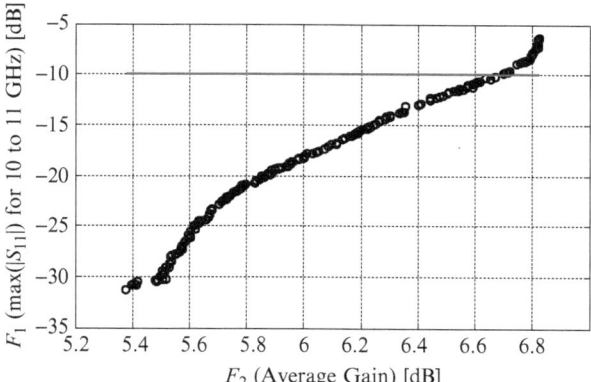

11.3.3   Planar Yagi Antenna: Results

Using the surrogate model created as described in Sect. 11.3.2, the Pareto front was generated by means of the multi-objective evolutionary algorithm (MOEA). Figure 11.9 shows the Pareto front obtained by applying the MOEA. It can be observed that the level of antenna reflection within the frequency band 10–11 GHz can change from around −9 dB to about −30 dB, whereas the average end-fire gain can change between 5.2 and 6.8 dB. Seventeen designs out of Pareto front have been refined at step 5 of the optimization algorithm (Sect. 11.1.2). These refined designs are listed in Table 11.2; see also Fig. 11.10. Responses of the selected designs are shown in Figs. 11.11 and 11.12. The results indicate the capability of the considered antenna structure in terms of matching the gain, as well as the possible trade-offs between these two quantities.

As mentioned before, the computational cost of creating the kriging surrogate corresponds to about 60 evaluations of the high-fidelity model R_f.

Multi-objective optimization of R_s takes about 20 min. Design refinement takes, on average, two iterations of the OSM algorithm (i.e., two evaluations of R_f). Thus, the total design cost (obtaining Pareto front and 17 antenna designs as in Table 11.1) corresponds to about 80 evaluations of the high-fidelity model.

11.4   Summary

In this chapter, multi-objective design optimization of antennas has been demonstrated. By using variable-fidelity EM simulation models as well as auxiliary response surface approximations (here, kriging), it is possible to obtain the Pareto

Table 11.2 Yagi antenna: selected optimization results

	Design variables (mm)								Max$|S_{11}|$ gain[a]
	s_1	s_2	v_1	v_2	u_1	u_2	u_3	u_4	(dB)
1	4.13	3.23	9.55	4.65	3.92	4.72	2.19	1.68	−24.8 5.2
2	4.16	3.38	9.34	4.67	3.9	4.76	2.2	1.72	−24 5.5
3	4.15	3.45	9.38	4.7	3.82	4.86	2.16	1.74	−23.2 5.5
4	4.15	3.55	9.29	4.68	3.87	4.8	2.21	1.73	−22 5.6
5	4.26	3.75	9.13	4.66	3.91	4.8	2.23	1.68	−20.9 5.7
6	4.22	3.82	9.13	4.68	3.87	4.83	2.22	1.68	−20 5.7
7	4.2	4.02	9.08	4.65	3.85	4.87	2.22	1.7	−18.9 5.8
8	4.33	4.21	8.77	4.65	3.81	4.86	2.16	1.62	−18.3 5.9
9	4.36	4.28	8.7	4.79	3.76	4.83	2.28	1.74	−17.2 6.1
10	4.3	4.25	8.76	4.77	3.88	4.84	2.21	1.65	−16 6.1
11	4.31	4.26	8.6	4.82	3.86	4.81	2.17	1.67	−15 6.2
12	4.31	4.26	8.57	4.87	3.9	4.82	2.17	1.64	−14 6.2
13	4.37	4.26	8.47	4.98	3.84	4.85	2.23	1.65	−13 6.3
14	4.36	4.22	8.46	5.03	3.85	4.86	2.2	1.63	−12 6.3
15	4.35	4.29	8.38	5.1	3.82	4.85	2.22	1.63	−11 6.5
16	4.34	4.25	8.26	5.12	3.92	4.75	2.2	1.65	−10.1 6.5
17	4.33	4.26	8.21	5.16	3.91	4.74	2.16	1.58	−9 6.5

[a]End-fire gain averaged over the 10–11 GHz bandwidth

Fig. 11.10 Pareto optimal designs obtained in the refinement stage. F_2 (Average gain) stands for the end-fire IEEE gain averaged over the 10–11 GHz band

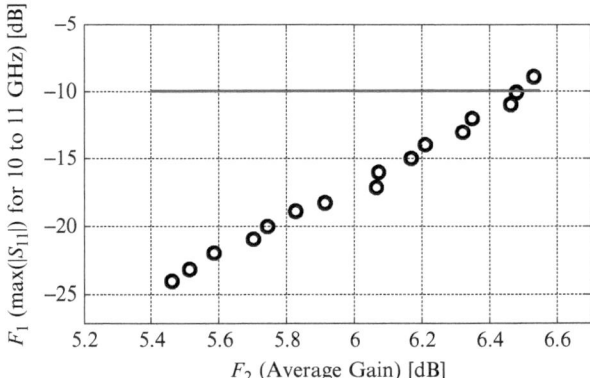

Fig. 11.11 Reflection responses of selected Pareto optimal designs (Fig. 11.10) obtained in the refinement stage: design 1 (*solid line*), design 6 (*dashed line*), design 11 (*dotted dashed line*), design 16 (*dotted line*)

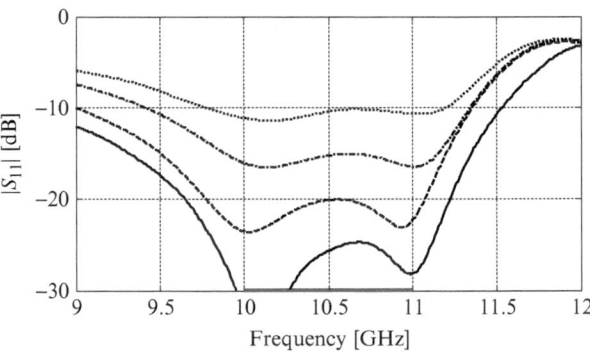

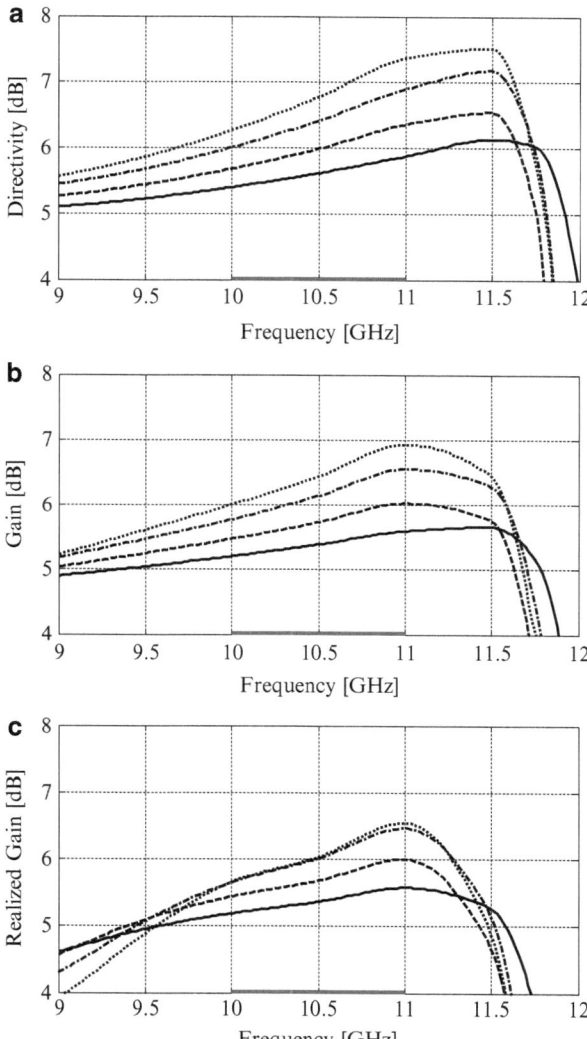

Fig. 11.12 End-fire radiation responses of selected Pareto optimal designs (Fig. 11.10) obtained in the refinement stage, design 1 (*solid line*), design 6 (*dashed line*), design 11 (*dotted dashed line*), and design 16 (*dotted line*): (**a**) directivity, (**b**) gain, (**c**) realized gain

front representing the performance trade-offs at a low computational cost. It should be emphasized that the design procedures discussed here utilize a number of concepts, including the aforementioned multilevel simulations, response surface approximations, and surrogate modeling, but also problem decomposition and multi-objective evolutionary algorithms.

Chapter 12
Practical Aspects of Surrogate-Based Antenna Design: Selecting Model Fidelity

As indicated in the previous chapters, successful design of antenna structures using surrogate-based optimization—within the class of techniques being the subject of this book—is a combination of a proper selection of a specific optimization technique and, perhaps even more importantly, a careful setup of the low-fidelity model. The construction of the low-fidelity antenna models has been elaborated in Chap. 5. While the overall goal is to have the low-fidelity model as fast as possible and, at the same time, as accurate as possible, these two characteristics are normally conflicting so that the computational speed has to be traded off for the accuracy of representing the high-fidelity model. In this chapter, we investigate in more details the importance of the low-fidelity model selection and its influence on the performance of the surrogate-based antenna optimization both in terms of the quality of the final solution and the overall design cost. Furthermore, we demonstrate that the use of multiple models of different fidelity may be beneficial to reduce the design cost while maintaining the robustness of the optimization process. Recommendations regarding the selection of the surrogate model coarseness are also given.

12.1 Selecting Model Fidelity: Speed Versus Accuracy Trade-Offs

The optimization methods considered in this book exploit an auxiliary low-fidelity model to construct the surrogate, the latter utilized as a predictor tool that leads us towards an improved antenna design. Computational cost and the accuracy of representing the high-fidelity model are the two most important characteristics of the low-fidelity model. As discussed in Chap. 5, the primary way of constructing the low-fidelity antenna models is through coarse-discretization EM simulation and/or applying other simplifications. While the lower-fidelity EM simulation is faster than the high-fidelity one, its cost cannot be neglected. Typically the time evaluation ratio between the high- and low-fidelity antenna models does not exceed 50, but in

S. Koziel and S. Ogurtsov, *Antenna Design by Simulation-Driven Optimization*, SpringerBriefs in Optimization, DOI 10.1007/978-3-319-04367-8_12, © Slawomir Koziel and Stanislav Ogurtsov 2014

some cases it can be as low as 5–10 (Koziel and Ogurtsov 2012b). In general, reducing the cost of the model is possible at the expense of its accuracy.

The speed and the accuracy of the low-fidelity model affect the performance of the surrogate-based optimization process. However, finding an optimum trade-off between the two factors is far from being obvious: coarser models are faster, which translates into lower cost per design iteration of the SBO algorithm. On the other hand, coarser models are also less accurate, which may results in a larger number of iterations necessary to yield a satisfactory design. Also, there is an increased risk that the optimization algorithm will fail to find a good design. Finer models, on the other hand, are more expensive, but they are more likely to produce a useful design in a smaller number of iteration. As indicated in Chap. 5, there is no fully automated way of selecting the right low-fidelity model for a given design problem and a given SBO method, and visual inspection of the model responses is still the most important criterion for making such a selection. In the remaining sections of this chapter, we try to highlight the importance of this problem using specific design cases.

12.2   Case Study 1: Design of Broadband Slot Antenna Using Output Space Mapping

Consider a CPW-fed slot antenna shown in Fig. 12.1a (Jiao et al. 2007). The design variables are $x = [a_x \, a_y \, a \, b \, s_1]^T$, $w_0 = 4$ mm, and $s_0 = 0.3$ mm. The substrate, 0.813 mm Rogers RO4003C ($\varepsilon_1 = 3.38$ at 10 GHz), and the ground plane are of infinite lateral extends. The initial design is $x^{(0)} = [40 \, 25 \, 10 \, 20 \, 2]^T$ mm. The design specifications are $|S_{11}| \leq -12$ dB for 2.3–7.6 GHz. The high-fidelity model R_f is evaluated with the CST MWS transient solver (CST MWS 2013) (3,556,224 mesh cells, simulated in 60 min). We consider three coarse models (all evaluated in CST MWS): R_{c1} (110,208 mesh cells, 1.5 min), R_{c2} (438,850, 5 min), and R_{c3} (1,113,840, 8 min).

Figure 12.1b shows the responses of R_f and R_{c1} through R_{c3} at the initial design. Because of mostly the vertical shift between the low- and the high-fidelity model responses, the surrogate model for the SBO algorithm (3.1) is created using output space mapping (OSM) (Bandler et al. 2004a, b) so that $R_s^{(i)}(x) = R_{ck}(x) + [R_f(x^{(i)}) - R_{ck}(x^{(i)})]$, k being an index of a respective low-fidelity model.

Table 12.1 and Fig. 12.1c show the optimization results. All the low-fidelity models are relatively reliable here and the qualities of the final designs are comparable. The design cost is the smallest for the SBO algorithm working with R_{c1} even though five design iterations are necessary. The algorithm working with R_{c2} and R_{c3} require only 3 and 2 iterations, respectively, but they are relatively expensive compared to R_f. Thus, in this case, using the coarsest model is the most advantageous.

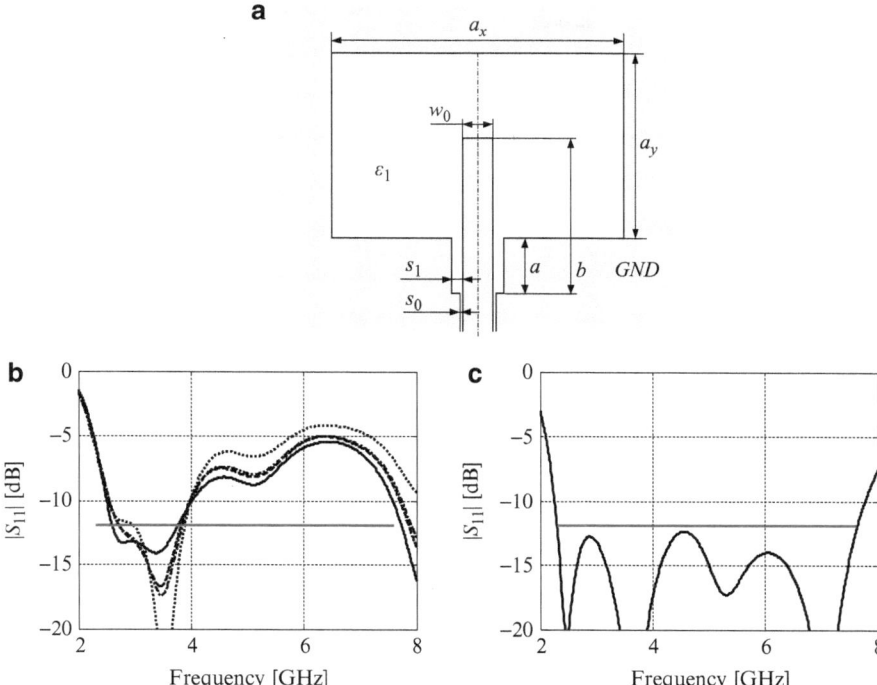

Fig. 12.1 CPW-fed broadband slot antenna: (**a**) geometry (Jiao et al. 2007); (**b**) model responses at the initial design, R_{c1} (*dotted line*), R_{c2} (*dotted dashed line*), R_{c3} (*dashed line*), and R_f (*solid line*); (**c**) high-fidelity model response at the final design found using the low-fidelity model R_{c3}

Table 12.1 CPW-fed slot antenna: design results

Low-fidelity model	Design cost: number of model evaluations[a]		Relative design cost[b]	Max$|S_{11}|$ for 2–8 GHz at final design (dB)
	R_c	R_f		
R_{c1}	287	5	12.2	−12.1
R_{c2}	159	3	16.2	−12.0
R_{c3}	107	2	16.3	−12.3

[a]Number of R_f evaluations is equal to the number of SBO iterations
[b]Equivalent number of R_f evaluations

12.3   Case Study 2: Model Management for Hybrid DRA

Consider a hybrid DRA shown in Fig. 12.2. The DRA is fed by a 50 Ω microstrip terminated with an open-ended section. Microstrip substrate is 0.787 mm thick Rogers RT5880. The design variables are $x = [h_0 \, r_1 \, h_1 \, u \, l_1 \, r_2]^T$. Other dimensions are fixed: $r_0 = 0.635$, $h_2 = 2$, $d = 1$, and $r_3 = 6$, all in mm. Permittivity of the DRA core is

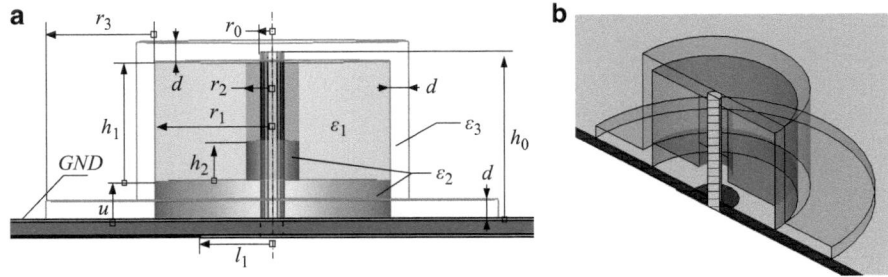

Fig. 12.2 Hybrid DRA: (**a**) side view, (**b**) 3D cut view (Koziel and Ogurtsov 2012b)

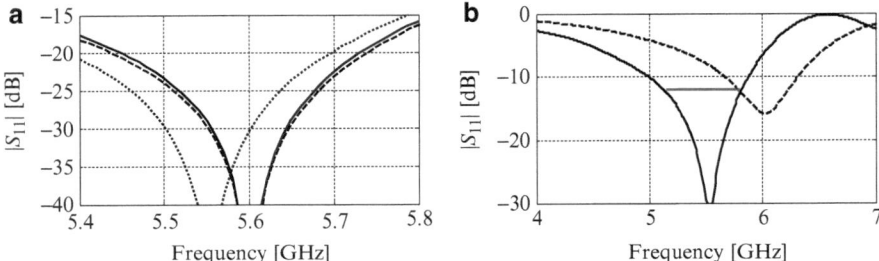

Fig. 12.3 Hybrid DRA: (**a**) high- (*solid line*) and low-fidelity model R_{c2} response at certain design before (*dotted line*) and after (*dashed line*) applying the frequency scaling, (**b**) high-fidelity model response at the initial design (*dashed line*) and at the final design obtained using the SBO algorithm using the low-fidelity model R_{c2} (*solid line*) (Koziel and Ogurtsov 2012b)

36, and the loss tangent is 10^{-4}, both at 10 GHz. The DRA support material is Teflon ($\varepsilon_2 = 2.1$), and the radome is of polycarbonate ($\varepsilon_3 = 2.7$ and $\tan\delta = 0.01$). The radius of the ground plane opening, shown in Fig. 12.3b, is 2 mm.

The high-fidelity antenna model $R_f(x)$ is evaluated using the time domain solver of CST Microwave Studio (CST MWS 2013) (~1,400,000 meshes, evaluation time 60 min). The goal is to adjust geometry parameters so that the following specifications are met: $|S_{11}| \le -12$ dB for 5.15–5.8 GHz. The initial design is $x^{(0)} = [7.0\ 7.0\ 5.0\ 2.0\ 2.0\ 2.0]^T$ mm.

We consider two auxiliary models of different fidelity, R_{c1} (~45,000 meshes, evaluation time 1 min) and R_{c2} (~300,000 meshes, evaluation time 3 min). We investigate the SBO algorithm (3.1) using either one of these models or both (R_{c1} at the initial state and R_{c2} in the later stages). The surrogate model is constructed using both output SM (cf. Sect. 12.2) and the frequency scaling (cf. (3.14), (3.15)). Figure 12.3a shows the importance of the frequency scaling, which, due to the shape similarity of the high- and low-fidelity model responses, allows substantial reduction of the misalignment between them.

The DRA design optimization has been performed three times: (1) the surrogate constructed using R_{c1} (cheaper but less accurate (Case 1)), (2) the surrogate

Table 12.2 Hybrid DRA design results (Koziel and Ogurtsov 2012b)

Case	Number of iterations	Number of model evaluations[a]			Total design cost[b]	Max$\|S_{11}\|$ for 5.15–5.8 GHz at final design (dB)
		R_{c1}	R_{c2}	R_f		
1	4	250	0	4	8.2	−12.6
2	2	0	150	2	9.5	−12.6
3	2	100	50	2	6.2	−12.6

[a]Number of R_f evaluations is equal to the number of SBO iterations
[b]Equivalent number of R_f evaluations

constructed using R_{c2} (more expensive but also more accurate (Case 2)), and (3) the surrogate constructed with R_{c1} at the first iteration and with R_{c2} for subsequent iterations (Case 3). The last option allows us to faster locate the approximate high-fidelity model optimum and then refine it using the more accurate model. The number of surrogate model evaluations was limited to 100 in the first iteration (as it involves the largest design change) and to 50 in the subsequent iterations (which requires smaller design modifications).

Table 12.2 shows the optimization results for all three cases. Figure 12.3b shows the high-fidelity model response at the final design obtained using the SBO algorithm working with low-fidelity model R_{c2}. The quality of the final designs found in all cases is the same. However, the SBO algorithm using the low-fidelity model R_{c1} (Case 1) requires more iterations than the algorithm using the model R_{c2} (Case 3), which is because the latter is more accurate. In this particular case, the overall computational cost of the design process is still lower for R_{c1} than for R_{c2}. On the other hand, the cheapest approach is Case 2 when the model R_{c1} is utilized in the first iteration that requires the largest number of EM analyses, whereas the algorithm switches to R_{c2} in the second iteration, which allows us to both reduce the number of iterations and number of evaluations of R_{c2} at the same time. The total design cost is the lowest overall.

12.4   Discussion and Recommendations

The numerical results presented in Sects. 12.2 and 12.3 as well as the results shown in Sect. 8.3, where a design of a wideband microstrip antenna using SBO and coarse-discretization EM models of various fidelities was investigated, allow us to draw some conclusions regarding the selection of the model fidelity for surrogate-based antenna optimization. While using the cheaper (and less accurate) model may translate into lower design cost, it also increases the risk of failure. Using the higher-fidelity model may increase the cost but it definitely improves the robustness of the SBO design process and reduces the number of iterations necessary to find a satisfactory design. Visual inspection of the low- and high-fidelity model responses remains—so far—the most important way of accessing the model quality, which may also give a hint which type of model correction should be applied while creating the surrogate.

The following rules of thumb and the "heuristic" model selection procedure can be formulated:

1. A parametric study of the low-fidelity model "coarseness" should be performed at the initial design in order to find the "coarsest" model that still adequately represents all the important features of the high-fidelity model response. The assessment should be done by visual inspection of the model responses having in mind that the critical factor is not the absolute model discrepancy but the similarity of the response shape (e.g., even relatively large frequency shift can be easily reduced by a proper frequency scaling).
2. When in doubt, it is safer to use a slightly finer low-fidelity model rather than a coarser one so that potential cost reduction is not lost due to a possible algorithm failure to find a satisfactory design.
3. The type of misalignment between the low- and high-fidelity model should be observed in order to properly select the type of low-fidelity model correction while constructing the surrogate. The two methods considered in the numerical examples presented in this chapter (i.e., the additive response correction and frequency scaling) can be considered as safe choices for many situations.

It should be emphasized that for some antenna structures, such as some narrowband antennas or wideband travelling wave antennas, it is possible to obtain quite good ratio between the simulation times of the high- and low-fidelity models (e.g., up to 50), which is because even for relatively coarse mesh, the low-fidelity model may still be a good representation of the high-fidelity one. For other structures (e.g., multi-resonant antennas), only much lower ratios (e.g., 5–10) may be possible, which would translate into lower design cost savings while using the surrogate-based optimization techniques.

Chapter 13
Discussion and Recommendations

Throughout this book, a number of simulation-driven optimization techniques suitable for the design of antenna structures have been discussed. Majority of the methods presented here are based on variable-fidelity EM simulations. The coarse-discretization EM simulations are used as an underlying low-fidelity model, which after suitable correction serves as a fast prediction tool leading to the improved design at a low computational cost. It has been demonstrated that the typical computational cost of the design process expressed in terms of the number of equivalent high-fidelity model evaluations is comparable to the number of design variables. In this chapter, we attempt to qualitatively compare these methods and give some recommendations for the readers interested in using them in their research and design work. Brief summaries of the SBO techniques considered in this book and their important properties are provided in Sect. 13.1. In Sect. 13.2 a set of recommendations and guidelines are formulated that might help in choosing the most suitable approach for a given antenna design problem.

13.1 SBO Methods Highlights

Most of the methods considered in this book use the similar SBO scheme (cf. 3.1), (Koziel et al. 2011c). The most important difference between them is in construction of the surrogate model. Simple techniques, such as basic response correction (see Sect. 4.2.3) or frequency scaling (cf. 3.14, 3.15), are very easy to implement but are not able to fully exploit the knowledge embedded in the low-fidelity model. More involved methods, such as SPRP (Koziel 2010a), usually offer better efficiency, however, at the expense of more complex implementation and certain assumptions which may or may not be satisfied in a given situation.

Space mapping (Sect. 4.2, Bandler et al. 2004a, b; Koziel et al. 2006) is a quite generic method. In particular, it is able to work even if the low-fidelity model is moderately accurate. In particular, by a suitable choice of the SM transformations,

S. Koziel and S. Ogurtsov, *Antenna Design by Simulation-Driven Optimization*, SpringerBriefs in Optimization, DOI 10.1007/978-3-319-04367-8_13, © Slawomir Koziel and Stanislav Ogurtsov 2014

the misalignment between the low- and the high-fidelity model can be reduced effectively. On the other hand, SM requires some experience in selecting the proper type of the surrogate model to be used (Koziel and Bandler 2007). As mentioned throughout this book, application of SM for antenna design usually requires an auxiliary response surface model (e.g., kriging interpolation one; Queipo et al. 2005) so that the computational overhead of the SM parameter extraction process is kept on acceptable levels.

Shape-preserving response prediction (SPRP, Sect. 4.3) can be extremely efficient because it exploits the knowledge embedded in the low-fidelity model to the fullest extent. However, it requires the response shape of the low- and high-fidelity models to be similar so that the characteristic points of these models are in one-to-one correspondence. Also, characteristic points of the SPRP model have to be selected in individual basis, which may require some experience. In this context, SPRP is the most suitable for antennas whose responses contain clear and distinctive features (e.g., narrowband and multiband antennas). With careful selection of the characteristic points, SPRP can be successfully applied for other types of problems including UWB antennas (cf. Sect. 6.2).

Adaptive response correction (ARC, Sect. 4.4; Koziel et al. 2009b) retains many features of SPRP without its limitations (restrictions on shape similarity between the low- and high-fidelity model responses). Its implementation is simpler than SPRP; however, it may not be as efficient as SPRP in "vertical" correction of the low-fidelity model.

Manifold mapping (MM, Sect. 4.5; Echeverria and Hemker 2005) is a very elegant and simple to implement generalization of the basic response correction technique (output SM). While it is not as efficient as SPRP or ARC, it can be conveniently combined with sensitivity data, which greatly improves its convergence properties (cf. Sect. 10.3).

Adaptively adjusted design specifications (AADS, Sect. 4.6; Koziel 2010b) is definitely the simplest method to implement. It does not require any correction of the low-fidelity model. Therefore, AADS can even be executed within any EM solver by modifying the design requirements and using built-in optimization capabilities. On the other hand, AADS only works with minimax-like design specifications. Also, AADS requires that the low-fidelity model is relatively accurate so that the possible discrepancies between the low- and high-fidelity models can be accounted for by design specifications adjustment.

Variable-fidelity simulation-driven optimization (VFSDO, Sect. 4.7; Koziel and Ogurtsov 2010b) is one of the most robust techniques, which is yet simple to implement. The only drawback is that it requires at least two low-fidelity models of different discretization density and some sort of initial study of the model accuracy and computational complexity trade-offs. While VFSDO will work with practically any setup, careful selection of mesh density can reduce the computational cost of the optimization process considerably.

As indicated in Chap. 10, derivative information, if available (e.g., through adjoint sensitivities; Nair and Webb 2003), can be utilized to improve properties of

Table 13.1 Qualitative comparison of simulation-driven surrogate-based techniques

Method	Complexity/ implementation	Issues/potential problems	Other comments
SM with Kriging (Sect. 4.2)	Moderate to complex	Requires user experience to select proper SM surrogate	More tolerant than AADS and SPRP in terms of low-fidelity model accuracy
SPRP (Sect. 4.3)	Moderate	Requires similar shape of low- and high-fidelity model responses $\geq$ may not work for antennas with complex/broadband responses	Selection of SPRP characteristic points has to be done individually for each design case
ARC (Sect. 4.4)	Moderate	Not as good as SPRP in "vertical" low-fidelity model correction	Retains features of SPRP without shape similarity requirements
MM (Sect. 4.5)	Easy	May oscillate near the optimum design due to the noisy low-fidelity model. Using with trust-region framework recommended	By definition utilizes high-fidelity data from multiple designs considered during the optimization run
AADS (Sect. 4.6)	Easy	Requires relatively accurate low-fidelity model; may not work for antennas with complex responses	Very convenient, can be executed within any EM solver using its built-in optimization capability
VFDSO (Sect. 4.7)	Easy to moderate	Meshing density for low-fidelity models has to be carefully selected to maintain good numerical efficiency	Requires two or more low-fidelity models; tolerant to low-fidelity model inaccuracy
Method using sensitivity information (Chap. 10)	Moderate	Requires cheap sensitivity information, normally obtained using adjoints. Availability depending on a specific EM solver	Can be combined with various SBO techniques. Improves robustness (better convergence) and reduces design cost

most SBO techniques both in terms of the robustness (particularly, convergence properties) and reducing computational cost of the design process. From this point of view, methods with sensitivity should not be considered as a separate category in the SBO group but rather as an enhancement technique.

Table 13.1 summarizes the main features of various SBO techniques discussed in the book.

13.2   Discussion and Recommendations

Selection of the most suitable SBO technique for a given antenna design task is an open problem in the sense that no rigorous procedures exist so far that would facilitate it. In general, the method selection should take into account the following factors:

- User experience regarding numerical modeling and optimization techniques (potentially less efficient but simpler techniques should be used for less-experienced users).
- Problem-specific knowledge and visual inspection of the antenna responses (visual assessment of the differences between the low- and high-fidelity model responses may help in determining the most suitable approach to construct the surrogate model).
- Available low-fidelity models (e.g., selection of the optimization technique may be limited if the low-fidelity model is relatively expensive compared to the high-fidelity one).

Having this in mind, both AADS and VFSDO techniques are preferred for the users with little or no optimization experience; however, AADS should be used if the low- and high-fidelity model responses are similar in shape. VFSDO is much more tolerant with this respect. SM with kriging, SPRP, and ARC are recommended for the users with more experience in optimization. In general, both SPRP and ARC are easier to implement and more efficient than SM; however, the latter is more tolerant to low-fidelity model inaccuracy. The examples presented in Chaps. 6–11 are representative in terms of the type of the antenna design problems that can be handled by the considered techniques. More information can be found in the literature. Interested reader is referred to the final sections of the respective chapters for specific references.

It should be emphasized that simple methods are often very efficient. The two basic techniques, i.e., additive response correction (cf. Sect. 4.2.3) and frequency scaling (cf. 3.14), are recommended if the major type of the discrepancies between the low- and high-fidelity models is vertical shift or frequency shift, respectively. The examples of using these basic methods have been discussed, among others, in Sect. 8.3 and Chap. 12.

In some situations, direct application of one "off-the-shelf" methods may not be as efficient as a customized approach, often combined with problem decomposition. This was demonstrated in Chap. 9 in case of antenna array optimization, as well as in Chap. 11 (multi-objective optimization).

The use of derivative information is recommended whenever available (particularly through adjoint sensitivity technique). As discussed in Chap. 10, sensitivity data can enhance many SBO techniques and even guarantee algorithm convergence in a classical sense (this is generally not ensured for SBO methods not using derivatives).

It should also be emphasized that selection and setup of the low-fidelity model is just as important step of the surrogate-based optimization process as the selection of the optimization methods itself. This topic has been discussed in detail in Chap. 12.

13.3 Prospective Look

A few words should be said about the perspectives of surrogate-based methods for antenna design, particularly about methods based on variable-fidelity EM simulations. As demonstrated in this book, variable-fidelity SBO methods can be extremely efficient, leading to satisfactory design at low computational cost corresponding to a handful of evaluations of the high-fidelity EM model of the antenna of interest. On the other hand, SBO methods are not as automated and therefore not as robust as many conventional techniques. There are open problems that include selection of the model fidelity, selection of the most appropriate way of constructing the surrogate model, and potential convergence issue. Successful application of these methods still partially depends on user experience. Research attempting to address these problems is ongoing. However appealing the SBO techniques are, it seems that their wide acceptance largely depends on implementing software tools where most of the decisions regarding model/method selection/setup could be done automatically without relying on the user interaction.

References

AEP: SMA Edge Mount P.C. Board Receptacles. Catalog. Applied Engineering Products, New Haven, CT, USA (2013). http://aepconnectors.com/pdf/SMAedg.pdf

Alexandrov, N.M., Dennis, J.E., Lewis, R.M., Torczon, V.: A trust region framework for managing use of approximation models in optimization. Struct. Multidisc. Optim. **15**, 16–23 (1998)

Alexandrov, N.M., Lewis, R.M.: An overview of first-order model management for engineering optimization. Optim. Eng. **2**, 413–430 (2001)

Amari, S., LeDrew, C., Menzel, W.: Space-mapping optimization of planar coupled-resonator microwave filters. IEEE Trans. Microw. Theory Tech. **54**, 2153–2159 (2006)

Angiulli, G., Cacciola, M., Versaci, M.: Microwave devices and antennas modelling by support vector regression machines. IEEE Trans. Magn. **43**, 1589–1592 (2007)

Ares-Pena, F.J., Rodriguez-Gonzalez, A., Villanueva-Lopez, E., Rengarajan, S.R.: Genetic algorithms in the design and optimization of antenna array patterns. IEEE Trans. Antennas Propag. **47**, 506–510 (1999)

Audet, C., Dennis Jr., J.E.: Mesh adaptive direct search algorithms for constrained optimization. SIAM J. Optim. **17**, 188–217 (2006)

Back, T., Fogel, D.B., Michalewicz, Z. (eds.): Evolutionary Computation 1: Basic Algorithms and Operators. Taylor & Francis Group, Bristol (2000)

Bakr, M.H., Bandler, J.W., Biernacki, R.M., Chen, S.H., Madsen, K.: A trust region aggressive space mapping algorithm for EM optimization. IEEE Trans. Microw. Theory Tech. **46**, 2412–2425 (1998)

Bakr, M.H., Bandler, J.W., Georgieva, N.K., Madsen, K.: A hybrid aggressive space-mapping algorithm for EM optimization. IEEE Trans. Microw. Theory Tech. **47**, 2440–2449 (1999)

Bakr, M.H., Ghassemi, M., Sangary, N.: Bandwidth enhancement of narrow band antennas exploiting adjoint-based geometry evolution. IEEE Int. Symp. Antennas Prop., pp. 2909–2911. Spokane, WA (2011). 3–8 July 2011

Balanis, C.A.: Antenna Theory, 3rd edn. Wiley-Interscience, Hoboken, NJ (2005)

Bandler, J.W., Seviora, R.E.: Wave sensitivities of networks. IEEE Trans. Microw. Theory Tech. **20**, 138–147 (1972)

Bandler, J.W., Biernacki, R.M., Chen, S.H., Swanson Jr., D.G., Ye, S.: Microstrip filter design using direct EM field simulation. IEEE Trans. Microw. Theory Tech. **42**, 1353–1359 (1994)

Bandler, J.W., Biernacki, R.M., Chen, S.H., Hemmers, R.H., Madsen, K.: Electromagnetic optimization exploiting aggressive space mapping. IEEE Trans. Microw. Theory Tech. **43**, 2874–2882 (1995)

Bandler, J.W., Cheng, Q.S., Gebre-Mariam, D.H., Madsen, K., Pedersen, F., Søndergaard, J.: EM-based surrogate modeling and design exploiting implicit, frequency and output space mappings. IEEE Int. Microw. Symp. Digest, pp. 1003–1006. Philadelphia, PA (2003)

Bandler, J.W., Cheng, Q.S., Dakroury, S.A., Mohamed, A.S., Bakr, M.H., Madsen, K., Søndergaard, J.: Space mapping: the state of the art. IEEE Trans. Microw. Theory Tech. **52**, 337–361 (2004a)

Bandler, J.W., Cheng, Q.S., Nikolova, N.K., Ismail, M.A.: Implicit space mapping optimization exploiting preassigned parameters. IEEE Trans. Microw. Theory Tech. **52**, 378–385 (2004b)

Basudhar, A., Dribusch, C., Lacaze, S., Missoum, S.: Constrained efficient global optimization with support vector machines. Struct. Multidisc. Optim. **46**, 201–221 (2012)

Beachkofski, B., Grandhi, R.: Improved distributed hypercube sampling. American Institute of Aeronautics and Astronautics. Paper AIAA 2002, p. 1274 (2002)

Bevelacqua, P.J., Balanis, C.A.: Optimizing antenna array geometry for interference suppression. IEEE Trans. Antennas Propag. **55**, 637–641 (2007)

Björkman, M., Holmström, K.: Global optimization of costly nonconvex functions using radial basis functions. Optim. Eng. **1**, 373–397 (2000)

Booker, A.J., Dennis, J.E., Frank, P.D., Serafini, D.B., Torczon, V., Trosset, M.W.: A rigorous framework for optimization of expensive functions by surrogates. Struct. Optim. **17**, 1–13 (1999)

Broyden, C.G.: A class of methods for solving nonlinear simultaneous equations. Math. Comp. **19**, 577–593 (1965)

Chen, Z.N.: Wideband microstrip antennas with sandwich substrate. IET Microw. Antennas Propag. **2**, 538–546 (2008)

Cheng, Q.S., Rautio, J.C., Bandler, J.W., Koziel, S.: Progress in simulator-based tuning—the art of tuning space mapping. IEEE Microw. Mag. **11**, 96–110 (2010)

Chung, Y.S., Cheon, C., Park, I.H., Hahn, S.Y.: Optimal design method for microwave device using time domain method and design sensitivity analysis-part II: FDTD case. IEEE Trans. Magn. **37**, 3255–3259 (2001)

Conn, A.R., Gould, N.I.M., Toint, P.L.: Trust Region Methods. MPS-SIAM Series on Optimization (2000)

Conn, A.R., Scheinberg, K., Vicente, L.N.: Introduction to Derivative-Free Optimization. MPS-SIAM Series on Optimization, MPS-SIAM (2009)

Couckuyt, I.: Forward and inverse surrogate modeling of computationally expensive problems. PhD Thesis, Ghent University (2013)

CST Microwave Studio: CST AG, Bad Nauheimer Str. 19, D-64289 Darmstadt, Germany (2013)

Deal, W.R., Kaneda, N., Sor, J., Qian, Y., Itoh, T.: A new quasi-Yagi antenna for planar active antenna arrays. IEEE Trans. Microw. Theory Tech. **48**, 910–918 (2000)

Deng, S.M., Tsai, C.L., Chiu, C.W., Chang, S.F.: CPW-fed dual rectangular ceramic dielectric resonator antennas through inductively coupled slots, in Proc. IEEE Antennas Propag. Soc. Int. Symp. **1**, 1102–1105 (2004)

Director, S.W., Rohrer, R.A.: The generalized adjoint network and network sensitivities. IEEE Trans. Circ. Theory **16**, 318–323 (1969)

Dorigo, M., Gambardella, L.M.: Ant colony system: a cooperative learning approach to the traveling salesman problem. IEEE Trans. Evol. Comput. **1**, 53–66 (1997)

Echeverria, D., Hemker, P.W.: Space mapping and defect correction. CMAM Int. Math. J. Comput. Methods Appl. Math. **5**, 107–136 (2005)

Echeverría, D.: Multi-Level optimization: space mapping and manifold mapping. Ph.D. Thesis, Faculty of Science, University of Amsterdam (2007a)

Echeverría, D.: Two new variants of the manifold-mapping technique. COMPEL. Int. J. Comput. Math. Electr. Eng. **26**, 334–344 (2007b)

Echeverría, D., Hemker, P.W.: Manifold mapping: a two-level optimization technique. Comput. Visual. Sci. **11**, 193–206 (2008)

El Sabbagh, M.A., Bakr, M.H., Nikolova, N.K.: Sensitivity analysis of the scattering parameters of microwave filters using the adjoint network method. Int. J. RF Microw. Comp. Aided Eng. **16**, 596–606 (2006)

Emmerich, M.T.M., Giannakoglou, K., Naujoks, B.: Single and multiobjective evolutionary optimization assisted by Gaussian random field metamodels. IEEE Trans. Evol. Comput. **10**, 421–439 (2006)

FEKO® User's Manual, Suite 6.0: EM Software & Systems-S.A. (Pty) Ltd, 32 Techno Lane, Technopark, Stellenbosch, 7600, South Africa (2011)

Fonseca, C.M.: Multiobjective genetic algorithms with applications to control engineering problems. PhD Thesis, Department of Automatic Control and Systems Engineering, University of Sheffield, Sheffield, UK (1995)

Forrester, A.I.J., Keane, A.J.: Recent advances in surrogate-based optimization. Prog. Aerosp. Sci. **45**, 50–79 (2009)

Giunta, A.A., Wojtkiewicz, S.F., Eldred, M.S.: Overview of modern design of experiments methods for computational simulations. American Institute of Aeronautics and Astronautics, Paper AIAA 2003, p. 0649 (2003)

Goldberg, D.E.: Genetic Algorithms in Search, Optimization & Machine Learning. Pearson Education, Singapore (1989)

Grimaccia, F., Mussetta, M., Zich, R.E.: Genetical swarm optimization: self-adaptive hybrid evolutionary algorithm for electromagnetics. IEEE Trans. Antennas Propag. **55**, 781–785 (2007)

Gunn, S.R.: Support vector machines for classification and regression. Technical Report. School of Electronics and Computer Science, University of Southampton (1998)

Gutmann, H.-M.: A radial basis function method for global optimization. J. Global Optim. **19**, 201–227 (2001)

Haupt, R.L.: Antenna design with a mixed integer genetic algorithm. IEEE Trans. Antennas Propag. **55**, 577–582 (2007)

Haykin, S.: Neural Networks: A Comprehensive Foundation, 2nd edn. Prentice Hall, Upper Saddle River, NJ (1998)

Harrington, R.F.: Field Computation by Moment Methods. Wiley-IEEE Press, Hoboken, NJ (1993)

Hemker, P.W., Echeverría, D.: A trust-region strategy for manifold mapping optimization. JCP J. Comput. Phys. **224**, 464–475 (2007)

HFSS: Release 13.0, ANSYS. http://www.ansoft.com/products/hf/hfss/ (2010)

Jacobs, J.P.: Bayesian support vector regression with automatic relevance determination kernel for modeling of antenna input characteristics. IEEE Trans. Antennas Propag. **60**, 2114–2118 (2012)

Jacobs, J.P., Koziel, S., Ogurtsov, S.: Computationally efficient multi-fidelity Bayesian support vector regression modeling of planar antenna input characteristics. IEEE Trans. Antennas Propag. **61**, 980–984 (2013)

Jacobson, P., Rylander, T.: Gradient-based shape optimization of conformal array antennas. IET Microw. Antennas Propag. **4**, 200–209 (2010)

Jiao, J.-J., Zhao, G., Zhang, F.-S., Yuan, H.-W., Jiao, Y.-C.: A broadband CPW-fed T-shape slot antenna. Progr. Electromag. Res. **76**, 237–242 (2007)

Jin, N., Rahmat-Samii, Y.: Advances in particle swarm optimization for antenna designs: real-number, binary, single-objective and multiobjective implementations. IEEE Trans. Antennas Propag. **55**, 556–567 (2007)

Jin, N., Rahmat-Samii, Y.: Analysis and particle swarm optimization of correlator antenna arrays for radio astronomy applications. IEEE Trans. Antennas Propag. **56**, 1269–1279 (2008)

Jin, Y.: Surrogate-assisted evolutionary computation: recent advances and future challenges. Swarm Evol. Comput. **1**, 61–70 (2011)

Jones, D., Schonlau, M., Welch, W.: Efficient global optimization of expensive black-box functions. J. Global Optim. **13**, 455–492 (1998)

Journel, A.G., Huijbregts, C.J.: Mining Geostatistics. Academic Press, London (1981)

Kabir, H., Wang, Y., Yu, M., Zhang, Q.J.: Neural network inverse modeling and applications to microwave filter design. IEEE Trans. Microw. Theory Tech. **56**, 867–879 (2008)

Kaneda, N., Deal, W.R., Qian, Y., Waterhouse, R., Itoh, T.: A broad-band planar quasi-Yagi antenna. IEEE Trans. Antennas Propag. **50**, 1158–1160 (2002)

Kazemi, M., Wang, G.G., Rahnamayan, S., Gupta, K.: Metamodelbased optimization for problems with expensive objective and constraint functions. ASME J. Mech. Des. **133**, 014505 (2011)

Kempel, L.C.: Computational electromagnetics for antennas. In: Volakis, J.L. (ed.) Antenna Engineering Handbook. McGraw Hill, New York (2007)

Kennedy, J.: The particle swarm: social adaptation of knowledge. Proc. 1997 Int. Conf. Evolutionary Computation, pp. 303–308. Indianapolis, IN (1997)

Kennedy, J., Eberhart, R.C., Shi, Y.: Swarm Intelligence. Academic Press, Boston, MA (2001)

Kirkpatrick, S., Gelatt, C.D., Vecchi, M.P.: Optimization by simulated annealing. Science **220**, 671–680 (1983)

Kishk, A.A., Antar, Y.M.: Dielectric resonator antennas. In: Volakis, J.L. (ed.) Antenna Engineering Handbook, 4th edn. McGraw Hill, New York (2007)

Kiziltas, G., Psychoudakis, D., Volakis, J.L., Kikuchi, N.: Topology design optimization of dielectric substrates for bandwidth improvement of a patch antenna. IEEE Trans. Antennas Propag. **51**, 2732–2743 (2003)

Kleijnen, J.P.C.: Kriging metamodeling in simulation: a review. Eur. J.Oper. Res. **192**, 707–716 (2009)

Koehler, J.R., Owen, A.B.: Computer experiments. In: Ghosh, S., Rao, C.R. (eds.) Handbook of Statistics, vol. 13, pp. 261–308. Elsevier Science B.V, The Netherlands (1996)

Kokotoff, D.M., Aberle, J.T., Waterhose, R.B.: Rigorous analysis of probe-fed printed annular ring antennas. IEEE Trans. Antennas Propag. **47**, 384–388 (1999)

Kolda, T.G., Lewis, R.M., Torczon, V.: Optimization by direct search: new perspectives on some classical and modern methods. SIAM Rev. **45**, 385–482 (2003)

Koulouridis, S., Psychoudakis, D., Volakis, J.: Multiobjective optimal antenna design based on volumetric material optimization. IEEE Trans. Antennas Propag. **55**, 594–603 (2007)

Koziel, S.: Surrogate-based optimization of microwave structures using space mapping and kriging. Proc. European Microwave Conference, Rome, Italy, 28 September to 2 October 2009, pp. 1062–1065 (2009)

Koziel, S.: Shape-preserving response prediction for microwave design optimization. IEEE Trans. Microw. Theory Tech. **58**, 2829–2837 (2010a)

Koziel, S.: Adaptively adjusted design specifications for efficient optimization of microwave structures. Prog. Electromag. Res. B (PIER B) **21**, 219–234 (2010b)

Koziel, S.: Efficient Optimization of Microwave Structures Through Design Specifications Adaptation. IEEE Int. Symp. Antennas Propag., Toronto, Canada (2010c)

Koziel, S.: Adaptive design specifications and coarsely-discretized EM models for rapid optimization of microwave structures. Appl. Comput. Electromag. Soc. J. **26**, 1007–1015 (2011)

Koziel, S., Bandler, J.W.: SMF: a user-friendly software engine for space-mapping-based engineering design optimization. Int. Symp. Signals, Systems and Electronics, URSI ISSSE 2007, Montreal, Canada, pp. 157–160 (2007a)

Koziel, S., Bandler, J.W.: Coarse and Surrogate Model Assessment for Engineering Design Optimization with Space Mapping, pp. 107–110. IEEE MTT-S Int. Microw. Symp. Dig., Honolulu, HI (2007b)

Koziel, S., Echeverría Ciaurri, D.: Reliable simulation-driven design optimization of microwave structures using manifold mapping. Prog. Electromag. Res. B **26**, 361–382 (2010)

Koziel, S., Ogurtsov, S.: Computationally efficient simulation-driven design of a printed 2.45 GHz yagi antenna. Microw. Opt. Technol. Lett. **52**, 1807–1810 (2010a)

Koziel, S., Ogurtsov, S.: Robust Multi-fidelity Simulation-Driven Design Optimization of Microwave Structures, pp. 201–204. IEEE MTT-S Int. Microw. Symp. Dig., Anaheim, CA (2010b)

Koziel, S., Ogurtsov, S.: Simulation-driven design in microwave engineering: methods. In: Koziel, S., Yang, X.S. (eds.) Computational Optimization, Methods and Algorithms, Series: Studies in Computational Intelligence. Springer-Verlag, Germany (2011a)

Koziel, S., Ogurtsov, S.: Simulation-driven design in microwave engineering: application case studies. In: Yang, X.S., Koziel, S. (eds.) Computational Optimization and Applications in Engineering and Industry, Series: Studies in Computational Intelligence. Springer-Verlag, Germany (2011b)

Koziel, S., Ogurtsov, S.: Rapid design optimization of antennas using space mapping and response surface approximation models. Int. J. RF Microw. CAE **21**, 611–621 (2011c)

Koziel, S., Ogurtsov, S.: Fast simulation-driven design of antennas using shape-preserving response prediction. IEEE Int. Symp. Antennas Propag., Spokane, WA, 2011, pp. 1338–1341 (2011d)

Koziel, S. Ogurtsov, S.: Rapid optimization of dielectric resonator antennas using surrogate models. Proc. Loughborough Antennas & Propagation Conference, LAPC 2011 (2011e)

Koziel, S., Ogurtsov, S.: Bandwidth Enhanced Design of Dielectric Resonator Antennas Using Surrogate-Based Optimization. IEEE Int. Symp. Antennas Propag., Spokane, WA (2011f)

Koziel, S., Ogurtsov, S.: Simulation-driven design of broadband antennas using surrogate-based optimization. In: Koziel, S., Zhang, X.S., Yang, Q.J. (eds.) Simulation-Driven Design Optimization and Modeling for Microwave Engineering. Imperial College Press, London (2012a)

Koziel, S., Ogurtsov, S.: Model management for cost-efficient surrogate-based optimization of antennas using variable-fidelity electromagnetic simulations. IET Microw. Antennas Propag. **6**, 1643–1650 (2012b)

Koziel, S., Ogurtsov, S.: Robust design of UWB antennas using response surface approximations and manifold mapping. 6th European Conference on Antennas and Propagation, EuCAP 2012, pp. 773–775 (2012c)

Koziel, S., Ogurtsov, S.: Reduced-cost design optimization of antenna structures using adjoint sensitivity. Microw. Opt. Technol. Lett. **54**, 2594–2597 (2012d)

Koziel, S., Ogurtsov, S.: EM-simulation-based antenna design using adaptive response correction. European Conf. Antennas & Propagation (2013a)

Koziel, S., Ogurtsov, S.: Computational-budget-driven automated microwave design optimization using variable-fidelity electromagnetic simulations. Int. J. RF Microw. CAE **22**, 349–356 (2013b)

Koziel, S., Ogurtsov, S.: Multi-level design optimization of microwave structures with automated model fidelity adjustment. IEEE Int. Microw. Symp., IMS 2013. Seattle, WA, USA (2013c)

Koziel, S., Ogurtsov, S.: Simulation driven design of a microstrip antenna array by means of surrogate-based optimization. Proc. EuCAP 2013, The 7th European Conference on Antennas and Propagation, 8–12 April 2013, Gothenburg, Sweden (2013d)

Koziel, S., Ogurtsov, S.: Design optimization of microstrip antenna arrays using surrogate-based methodology. IEEE Intl. Symposium on Antennas and Propagation, Orlando, Florida, USA, 7–13 July (2013e)

Koziel, S., Ogurtsov, S.: Shape-preserving response prediction with adjoint sensitivities for microwave design optimization. IEEE Int. Microw. Symp., IMS 2013. Seattle, WA, USA (2013f).

Koziel, S. Yang, X.S. (eds.) Computational optimization, methods and algorithms. Series: Studies in Computational Intelligence, vol. 356, Springer, Germany (2011)

Koziel, S., Bandler, J.W., Madsen, K.: Space mapping framework for engineering optimization: theory and implementation. IEEE Trans. Microw. Theory Tech. **54**, 3721–3730 (2006)

Koziel, S., Bandler, J.W., Madsen, K.: Quality assessment of coarse models and surrogates for space mapping optimization. Optim Eng. **9**, 375–391 (2008a)

Koziel, S., Cheng, Q.S., Bandler, J.W.: Space mapping. IEEE Microw. Mag. **9**, 105–122 (2008b)

Koziel, S., Meng, J., Bandler, J.W., Bakr, M.H., Cheng, Q.S.: Accelerated microwave design optimization with tuning space mapping. IEEE Trans. Microw. Theory Tech. **57**, 383–394 (2009a)

Koziel, S., Bandler, J.W., Madsen, K.: Space mapping with adaptive response correction for microwave design optimization. IEEE Trans. Microw. Theory Tech. **57**, 478–486 (2009b)

Koziel, S., Bandler, J.W., Cheng, Q.S.: Robust trust-region space-mapping algorithms for microwave design optimization. IEEE Trans. Microw. Theory Tech. **58**, 2166–2174 (2010a)

Koziel, S., Cheng, Q.S., Bandler, J.W.: Implicit space mapping with adaptive selection of preassigned parameters. IET Microw. Antennas Propag. **4**, 361–373 (2010b)

Koziel, S., Bandler, J.W., Cheng, Q.S.: Constrained parameter extraction for microwave design optimization using implicit space mapping. IET Microw. Antennas Propag. **5**, 1156–1163 (2011a)

Koziel, S., Ogurtsov, S., Bakr, M.H.: Computationally efficient design optimization of wideband planar antennas using cauchy approximation and space mapping. Microw. Opt. Technol. Lett. **53**, 618–622 (2011c)

Koziel, S., Mosler, F., Reitzinger, S., Thoma, P.: Robust microwave design optimization using adjoint sensitivity and trust regions. Int. J. RF Microw. CAE **22**, 10–19 (2012a)

Koziel, S., Ogurtsov, S., Szczepanski, S.: Rapid antenna design optimization using shape-preserving response prediction. Bull. Pol. Acad. Sci. Tech. Sci. **60**, 143–149 (2012b)

Koziel, S., Ogurtsov, S., Bandler, J.W., Cheng, Q.S.: Robust space mapping optimization exploiting EM-based models with adjoint sensitivity. IEEE MTT-S Int. Microwave Symp. Dig. (2012c)

Koziel, S., Ogurtsov, S., Bakr, M.H.: Antenna modeling using space-mapping corrected Cauchy-approximation surrogates. Microw. Opt. Technol. Lett. **54**, 37–40 (2012d)

Koziel, S., Ogurtsov, S., Couckuyt, I., Dhaene, T.: Variable-fidelity electromagnetic simulations and co-kriging for accurate modeling of antennas. IEEE Trans. Antennas Propag. **61**, 1301–1308 (2013)

Kuwahara, Y.: Multiobjective optimization design of yagi–uda antenna. IEEE Trans. Antennas Propag. **53**, 1984–1992 (2005)

Li, W.T., Shi, X.W., Hei, Y.Q., Liu, S.F., Zhu, J.: A hybrid optimization algorithm and its application for conformal array pattern synthesis. IEEE Trans. Antennas Propag. **58**, 3401–3406 (2008)

Lin, J.-M.: The Finite Element Method in Electromagnetics, 2nd edn. Wiley-IEEE Press, New York (2002)

Loshchilov, I., Schoenauer, M., Sebag, M.: Self-adaptive surrogate-assisted covariance matrix adaptation evolution strategy. Proceedings of the Genetic and Evolutionary Computation Conference (GECCO 2012), pp. 321–328 (2012)

Makarov, S.: Antenna and EM Modeling with Matlab. Wiley-Interscience, New York (2002)

Matlab ver. 7.14: Mathworks Inc., Natick, MA 01760, USA (2012)

Minsky, M.I., Papert, S.A.: Perceptrons: An Introduction to Computational Geometry. The MIT Press, Cambridge, MA (1969)

Nair, D., Webb, J.P.: Optimization of microwave devices using 3-D finite elements and the design sensitivity of the frequency response. IEEE Trans. Magn. **39**(3), 1325–1328 (2003)

Nocedal, J., Wright, S.J.: Numerical Optimization, Springer Series in Operations Research. Springer, New York (2000)

Ogurtsov, S., Koziel, S.: Rapid surrogate-based optimization of UWB planar antennas. Proc. 4th European Conf. Antennas Propag. (EuCAP 2010), pp. 1–4 (2010)

Ogurtsov, S., Koziel, S.: Optimization of UWB planar antennas using adaptive design specifications. Proc. 5th European Conf. Antennas Propag. (EuCAP 2011), pp. 2091–2094 (2011a).

Ogurtsov, S., Koziel, S.: Simulation-driven design of dielectric resonator antenna with reduced board noise emission. IEEE MTT-S Int. Microwave Symp. Dig. (2011b)

Ogurtsov, S., Koziel, S.: Design optimization of a dielectric ring resonator antenna for matched operation in two installation scenarios. Proc. International Review of Progress in Applied Computational Electromagnetics, ACES 2011, 27–31 March 2011, Williamsburg, VA, pp. 424–428 (2011c)

O'Hagan, A.: Curve fitting and optimal design for predictions. J. R. Statist. Soc. B **40**, 1–42 (1978)

Ong, Y.S., Nair, P.B., Keane, A.J.: Evolutionary optimization of computationally expensive problems via surrogate modeling. AIAA J. **41**, 687–696 (2003)

Pantoja, M.F., Meincke, P., Bretones, A.R.: A hybrid genetic algorithm space-mapping tool for the optimization of antennas. IEEE Trans. Antennas Propag. **55**, 777–781 (2007)

Parno, M.D., Hemker, T., Fowler, K.R.: Applicability of surrogates to improve efficiency of particle swarm optimization for simulation-based problems. Eng. Optim. **44**, 521–535 (2012)

Petko, J.S., Werner, D.H.: An autopolyploidy-based genetic algorithm for enhanced evolution of linear polyfractal arrays. IEEE Trans. Antennas Propag. **55**, 583–593 (2007)

Petosa, A.: Dielectric Resonator Antenna Handbook. Artech House, Atlanta, GA (2007)

Queipo, N.V., Haftka, R.T., Shyy, W., Goel, T., Vaidynathan, R., Tucker, P.K.: Surrogate-based analysis and optimization. Prog. Aerosp. Sci. **41**, 1–28 (2005)

Qing, X.M., Chen, Z.N.: Antipodal Vivaldi antenna for UWB applications. European Electromag. Symp., UWB SP 7 (2004)

Rasmussen, C.E., Williams, C.K.I.: Gaussian Processes for Machine Learning. MIT Press, Cambridge, MA (2006)

Rajo-Iglesias, E., Quevedo-Teruel, O.: Linear array synthesis using an ant-colony-optimization-based algorithm. IEEE Antennas Propag. Mag. **42**, 70–79 (2007)

Rayas-Sánchez, J.E.: EM-based optimization of microwave circuits using artificial neural networks: the state of the art. IEEE Trans. Microw. Theory Tech. **52**, 420–435 (2004)

Regis, R.G.: Evolutionary programming for high-dimensional constrained expensive black-box optimization using radial basis functions. IEEE Trans. Evol. Comput. pp. 1 (2013a)

Regis, R.G.: Constrained optimization by radial basis function interpolation for high-dimensional expensive black-box problems with infeasible initial points. Eng. Optim. **46**, 218–243 (2013b)

Roy, G.G., Das, S., Chakraborty, P., Suganthan, P.N.: Design of non-uniform circular antenna arrays using a modified invasive weed optimization algorithm. IEEE Trans. Antennas Propag. **59**, 110–118 (2011)

RO4000 Series High Frequency Circuit Materials. Rogers Corporation, Publication #92-004 (2010)

RT/duroid 6006/6010 Laminate: Data Sheet, Rogers Corporation. Advanced Circuit Materials Division, Chandler, AZ (2011)

Santner, T.J., Williams, B., Notz, W.: The Design and Analysis of Computer Experiments. Springer-Verlag, New York (2003)

Schantz, H.: The Art and Science of Ultrawideband Antennas. Artech House, Atlanta, GA (2005)

Selleri, S., Mussetta, M., Pirinoli, P., Zich, R.E., Matekovits, L.: Differentiated meta-PSO methods for array optimization. IEEE Trans. Antennas Propag. **56**, 67–75 (2008)

Shaker, G.S.A., Bakr, M.H., Sangary, N., Safavi-Naeini, S.: Accelerated antenna design methodology exploiting parameterized Cauchy models. J. Prog. Electromag. Res. (PIER B) **18**, 279–309 (2009)

Simpson, T.W., Peplinski, J., Koch, P.N., Allen, J.K.: Metamodels for computer-based engineering design: survey and recommendations. Eng. Comp. **17**, 129–150 (2001)

Smola, A.J., Schölkopf, B.: A tutorial on support vector regression. Statist. Comput. **14**, 199–222 (2004)

Special issue on synthesis and optimization techniques in electromagnetic and antenna system design. IEEE Trans. Antennas Propag., 55(3), part I, pp. 518–785 (2007)

Storn, R., Price, K.: Differential evolution—a simple and efficient heuristic for global optimization over continuous spaces. J. Global Optim. **11**, 341–359 (1997)

Shum, S., Luk, K.: Stacked anunular ring dielectric resonator antenna excited by axi-symmetric coaxial probe. IEEE Trans. Microw. Theory Tech. **43**, 889–892 (1995)

Taflove, A., Hagness, S.C.: Computational Electrodynamics: The Finite-Difference Time-Domain Method, 3rd edn. Artech House, Atlanta, GA (2006)

Toivanen, J.I., Makinen, R.A.E., Jarvenpaa, S., Yla-Oijala, P., Rahola, J.: Electromagnetic sensitivity analysis and shape optimization using method of moments and automatic differentiation. IEEE Trans. Antennas Propag. **57**, 168–175 (2009)

Torczon, W.: On the convergence of pattern search algorithms. SIAM J. Optim. **7**, 1–25 (1997)

Uchida, N., Nishiwaki, S., Izui, K., Yoshimura, M., Nomura, T., Sato, K.: Simultaneous shape and topology optimization for the design of patch antennas. European Conf. Antennas Prop., pp. 103–107 (2009)

Villemonteix, J., Vazquez, E., Walter, E.: An informational approach to the global optimization of expensive-to-evaluate functions. J. Global Optim. **44**, 509–534 (2009)

Volakis, J.L.: Antenna Engineering Handbook, 4th edn. McGraw-Hill, New York (2007)

Wild, S.M., Regis, R.G., Shoemaker, C.A.: ORBIT: optimization by radial basis function interpolation in trust-regions. SIAM J. Sci. Comput. **30**, 3197–3219 (2008)

Wu, K.-L., Zhao, Y.-J., Wang, J., Cheng, M.K.K.: An effective dynamic coarse model for optimization design of LTCC RF circuits with aggressive space mapping. IEEE Trans. Microw. Theory Tech. **52**, 393–402 (2004)

Yang, X.S.: Engineering Optimization: An Introduction with Metaheuristic Applications. Wiley, Hoboken, NJ (2010)

Zhang, Y., Pimpale, A., Meshram, M.K., Nikolova, N.K.: Printed antenna design using sensitivity analysis based on method of moment solution. IEEE Radio and Wireless Symposium, pp. 47–50 (2012)

Zhou, Z., Ong, Y.S., Nair, P.B., Keane, A.J., Lum, K.Y.: Combining global and local surrogate models to accelerate evolutionary optimization. IEEE Trans. Syst. Man Cybern. C Appl. Rev. **37**, 66–76 (2007)

Index

S. Koziel and S. Ogurtsov, *Antenna Design by Simulation-Driven Optimization*,
SpringerBriefs in Optimization, DOI 10.1007/978-3-319-04367-8,
© Slawomir Koziel and Stanislav Ogurtsov 2014